The Birth of the Moon

The
BIRTH OF THE MOON

by

Lewis A. Manson

ILLUSTRATED BY *J. NOEL*

Published in cooperation with
NORTHROP UNIVERSITY
Inglewood, California

DENNIS-LANDMAN publishers * Santa Monica, California

ACKNOWLEDGEMENTS

I am very grateful to Dr. Ulric B. Bray of Pasadena, California, Chairman of the Board of Trustees of Northrop University whose keen interest in this work led to its publication jointly by Dennis-Landman Publishers and Northrop University.

To Joseph J. Dailey, Jr., whose advice and help coupled with the great insight of J. Noel Dennis made this publication possible.

To my wife, Anne, who sacrificed much during the writing of these works.

Lewis A. Manson

IN MEMORIUM

Dr. Ulric B. Bray

September 10, 1900 - November 20, 1977

Printed in the United States of America
Library of Congress Catalog Card Number 76-19964

First printing

ISBN 0-930422-12-0

Contents

CREDITS

Photographs in this book are by the courtesy of:

Hale Observatories
Lick Observatory
Lockheed Solar Observatory
United States Department of the Interior
NASA

Color photographs of rock and mineral samples:

Joseph J. Dailey Jr. (many of the samples are from his personal collection)

Maps of the Moon:

USAF
U.S. Government Printing Office

Introduction

Many years ago, Dr. Robert S. Richardson, the great astronomer wisely told us that a very important, and perhaps the most important, ingredient in any scientific research is a very special imagination. As the reader peruses this book it will be obvious to him that the author has a super special insight, not in the sense of fantasy, but insight supported by a sufficient number of established facts combined with a large sprinkle of postulations that border on fact, opens the mind of the author in such a lucid manner that it makes it possible for the reader to see it all with his own eyes.

I had the pleasure to review one of the author's books, *Out of the Grey Mist.* This was back in 1955, long before man landed on the moon. The author made many predictions which disagreed with the most accepted theory of what we might find on the moon. Most of his theories were found to be true when man arrived on the moon to make personal observations.

Until now so many more questions have remained unanswered. In fact, the most important questions have remained unanswered. Questions such as: Where did the moon come from? Where did other planets come from? Why does Saturn have rings? And what happened to the atmosphere of Mars? Why do the different planets have different atmospheres? Why does our earth have a tilt to its axis? Why did the giant reptiles die so suddenly? How were coal and petroleum really formed?

I know you will enjoy reading the author's answers to these and many many more questions. In fact, you will be fascinated page by page.

John J. McKetta
The E.P. Schoch Professor
The University of Texas
Austin, Texas

Foreword

It is extremely disconcerting to look into the night sky and realize the Moon is a huge crustal section of the Earth, strewn with organic debris. The Earth paid the price for a near encounter with an interstellar body no more than an instant ago in geological time.

The *Birth of the Moon* is a most unique scientific piece of literature, relating the incredible details of a great cataclysm which overtook the Planet Earth. Astounding proof of the mechanics of this cataclysm is not only intrinsic to the logic offered by the material presented herein, but proof is also established in moon photographs supplied through NASA and the Jet Propulsion Laboratory, and in moon crater analyses furnished by the author. Geological mysteries are unlocked. New geological keys are presented which not only enable the geologist to unravel undecoded riddles, but also enable the uninitiated to consider the validity of catastrophism against the unreliability of uniformitarianism.

The book is the product of the imagination of one man, who has devoted his entire life to satisfy a compulsion to seek truth in the majesty of creation. Lewis Auman Manson spent his entire early life as an admirer of the vividness and beauty of nature and the natural causes of things. By the time he was fifteen, he was collecting and analyzing rock specimens from the gravel pits of Indiana. By the year 1943, at the age of 25, Lewis Manson had consigned himself to this inner compulsion to expose nature's most baffling enigmas. The author became more and more influenced by travels, from his early trips to gravel pits to later visits to the more awesome and fractured-fit-together mountains and gorges of the North, Central and South American continents.

Isolated thoughts and ideas would occur to the author at random. The Moon's birth, for instance, was formulated in the period 1943 to 1949 in a series of notes. These notes were scribbled and scratched, at times frantically as they occurred, on paper napkins and towels, old railroad receipts, calling cards, and any other shreds of momentarily available writing material. By 1950, after having accumulated literally hundreds of these notes, the ideas began to fit into a mosaic of interrelated events comprising a manuscript growing in size . . . It was about this time Lewis Manson realized that he was sketching a story of the universe which reached from "The Beginning" of creation, through a series of interlocking cosmic events, to the physical world we see about us today.

The manuscript was given the name *Out of the Grey Mist* and from the period 1950 to 1976 was brought to its present condition, a refined work built on the principle "a little here and a little there." A new theory of primal energy physics evolved, treating the atom as a universe unto itself, filled with systems whose ordered function is to act as the atomic building blocks of the universe. New theories unfolded describing the composition of the Sun, the cycle mechanics of the stars, and the geology of the Earth, and the "Great Catastrophe." In developing this work, it was the author's fondest desire to build on the basis of fact and adherence to the natural laws laid down by the Master Designer.

The *Birth of the Moon* is but one chapter from the larger volume work. It was selected by the author for publication because of its message and its relevancy to the world's present problems. It is believed that its contents constitute a chain of unbroken events which will create a new atmosphere of learning and a period of readjustment in a number of the physical sciences.

If indeed Mr. Lewis Manson's conceptual image is correct, and a large storehouse of evidence is in plain view, then truly, we, as parents of the new generation of children, the inheritors of this geological world we see about us, must encourage and support our children in their search for truth, awakening them to the divineness of the design of our universal home, the Planet Earth, divineness in the mysterious complexities which allows this globe to nurture and to shelter four billion human beings, each being capable of reaching into the spiritual resources of the stellar consciousness.

Joseph J. Dailey, Jr.
Gulf Oil Chemicals Company

Preface

In pictographs, clay tablets, ancient scrolls and books over the past thousands of years, numerous references have been made to catastrophes that have befallen the Earth and its peoples, but present geologists and other scientific investigators are not satisfied that any real evidence has ever been found in the deposits of till, sediments and rock layers of the Earth to give any scientific basis for these numerous references of world-wide catastrophes. These references have, therefore, been relegated to the world of myths and folk tales. Could it be that scholars and scientists have been looking in the wrong direction and at the wrong things, and basing their conclusions on the wrong basic premises?

The geomorphologist, A.K. Lobeck* refers to the French naturalist, George Cuvier (1769-1832), "Who believed that some sudden catastrophe befell the surface of the Earth some five or six thousand years ago, whereby the countries inhabited by man were devastated and their inhabitants were destroyed. At this time, he thought, there were great inundations by the sea and then a retreat of the waters:

> "It left behind also in northern countries, the carcasses of the great quadrupeds which are found embedded in the ice and preserved down to the present day intact with their hair, hides and flesh. It was at one and the same instant that these animals perished and that the glacial conditions came into existence.
>
> This change was sudden, instantaneous not gradual, and that which is so clearly the case in this last catastrophe is not less true of those which preceded it. The dislocation and overturning of the older strata show without any doubt that the causes which brought them into the position which they now occupy, were sudden and violent; and in like manner testimony to the violence of the movements which in many localities are found intercalated between beds of solid rocks.
>
> Life upon the earth in those times was often overtaken by these frightful occurrences. Living things without number were swept out of existence by catastrophes. Those inhabiting the dry lands were engulfed by deluge; others whose home was in the water perished when the sea bottom suddenly became dry land; whole races were extinguished leaving mere traces of their existence, which are now difficult to recognize, even by the naturalists. The evidence of those great and terrible events are everywhere to be clearly seen by anyone who knows how to read the record of the rocks."

* A.K. Lobeck, GEOMORPHOLOGY, AN INTRODUCTION TO THE STUDY OF LANDSCAPES, (McGraw-Hill Book Company 1939)

In preparing and writing the material in *Out of the Grey Mist* (the main work) and this book, I have adhered carefully to the following scientific method as well stated by A.K. Lobeck.*

FIRST STEP: OBSERVING THE FACTS

The first step always consists in the securing of a certain amount of data.

SECOND STEP: FORMULATION OF HYPOTHESES TO EXPLAIN THE OBSERVED FACTS

The general, separate, disconnected facts are then assembled with a view to discovering some general principle which will tie them together.

THIRD STEP: THE DEDUCTION OF FURTHER FACTS

Each hypothesis is next examined and studied as an abstract idea, with a view to determining what additional consequences must result if it be true. This procedure, the working from general to particular, is a deductive process of logic.

The importance of developing more than one hypothesis is readily comprehended when it is realized that with only a single hypothesis to defend, the investigator is prone to believe that it be the true one, he is prejudiced then in his search for further facts and is in danger of noting only those which support his hypothesis, to the conclusion of others.

FOURTH STEP: TESTING THE HYPOTHESIS AND ELIMINATING THE INVALID ONES

The final step consists in finding out which of the deduced facts are actually true. The false hypotheses are one by one discarded until there is but one single remaining hypothesis which stands up under the test. If the proof is sufficient, this hypothesis may then rank as a theory, or even as an established fact.

The writing of the main work, *Out of the Grey Mist,* of which this book is an extract, began in 1943 and at the completion of it, which required 33 years, most of the established ideas and theories still persist in the scientific world.

For example: the September, 1975 issue of Scientific American covered the solar system with a number of articles by various authors on subjects such as the Sun, planets, asteroids, etc. . . An article titled "The Origin and Evolution of the Solar System" presents the currently popular idea of the origin of all the bodies in the solar system from an accumulation of "dust and gases" that formed the Sun and the planets from one of the "curved arms of our spiral galaxy" and the article further states that some version of this theory is accepted by most astronomers. And in the same paragraph, the author states, "We still have no generally accepted theory to explain how the primitive solar nebula formed, how and when the Sun began to shine and how and when the planets coalesced out of the swirling dust." It is also part of this same theory that all of the stars in space formed out of dust and gases which accumulated, grew hot and began to shine.

Consider the following: First, no curved arm of our galaxy has yet been detected. Second, where did the gas and dust originate to form all of the solar bodies in space? The present theory assumes that the elements existed in dust and gas throughout space but gives no theory as to how these elements formed, or where they came from, to build the universe as we see it today.

* A.K. Lobeck, GEOMORPHOLOGY, AN INTRODUCTION TO THE STUDY OF LANDSCAPES, (McGraw-Hill Book Company 1939)

The story of "how it all began" is a fascinating one, and if you will follow this history I am about to relate to you, step by step, I promise that when you again look at any part of nature, whether the sky full of stars, a mountain, a rock, or even a handful of sand – in fact the whole natural world will be more meaningful to you.

In order to clearly understand how certain events came to pass in our physical universe, why certain phenomena exist, and why our immediate environment appears as it does, certain fundamental premises must be presented in order for one to understand all the processes and agencies that were at play during the activity described in this book.

Every single thing that exists is but one link in an endless chain of cause and effect. In telling the story of *The Birth of the Moon,* I have tried to build its chain of events without skipping any links, or without distorting any to make them fit, and most important of all – without using old links, or concepts, no longer valid.

THE MOON

Fig. 1.

USAF

Chapter 1

PROLOGUE

In spite of the great advancements in science in the last few years, advancements in nuclear physics, astronomy, geology and other sciences, the most important questions still remain unanswered. Questions such as: What is light? What is electricity? What is gravity? What is magnetism? Even the most fundamental question: How did it all begin? is still in the debate stage. None of the authorities in the related scientific disciplines agree on how, when or why our Universe came into being. There are three general theories: The Steady State, The Big Bang, and The Pulsating Theory. Each of these concepts has the backing of respected groups of scientists – and the controversy continues.

What are the Sun and stars made of and what keeps them burning? Why are the planets different from each other – some larger, some smaller, and of different densities and materials? Why do some planets have satellites and others do not? Where and how did Saturn acquire its rings and what happened to Pluto to make it the anomaly of the solar system? What happened to the atmosphere of Mars? Why does Venus have so much atmosphere? Why is its rotation retrograde to that of the Earth and the other major planets? Where do comets come from? What created the asteroids?

And our Earth. Why does it have the interesting 23½ degree tilt to its axis? Has it always had this tilt? Has the Earth always been the same distance from the Sun and has the Sun always risen in the east? Has the Earth always been the same size? Why has there been evidence of "polar shift"? Why has there been evidence of extreme climatic changes over the entire globe? Have the continents and the oceans remained much the same throughout the history of the Earth or have they changed their positions in respect to the equator and the poles? Or are they of recent appearance? What created the extensive fracture system found throughout the Earth?

How were the mountains formed? And why is the basement rock of the continents different from the basement rock of the oceans? If the mountains of the past were eroded and the material washed into the oceans (generally accepted opinion), where is all that material now?

What happened to all the animals at the time of the "great dying"? What caused this phenomenon, this sudden extinction of the giant reptiles? For eons the four corners of the Earth teemed with them, then suddenly they were gone, leaving only great beds of their bones and nests of their unhatched eggs. Why is evidence of ancient man extremely rare?

How was coal and petroleum formed? How was solid rock fractured in thousands of places then shot with veins of gold, silver and other metals? Why are vast areas of the Earth stained rich red by iron, black by maganese, and others deep purple by bauxite? Where and how were diamonds formed, and why are they found in deep pipes of strange clay?

Where did all the sand come from and all the pebbles and gravel? Why are level beds of sea shells found at the top of Mount Everest and tropical plants found in Spitsbergen and Antarctica? What causes volcanoes and why is lava from them so fertile to plant growth? Did water carve the Grand Canyon? Did ice carve the Yosemite Valley? What unique process formed the many caves found throughout the world? What really causes snow? rain? lightning? and the winds? What causes hurricanes, tornadoes and cyclones? What causes earthquakes?

And where did the Moon come from? What is it made of? Why does it look as it does? Is, or was there life on the moon?

Now – let us answer some of the questions. The following answers are based solely upon my personal research covering some 33 years. These are my ideas, theories, discoveries and works.

The Solar System

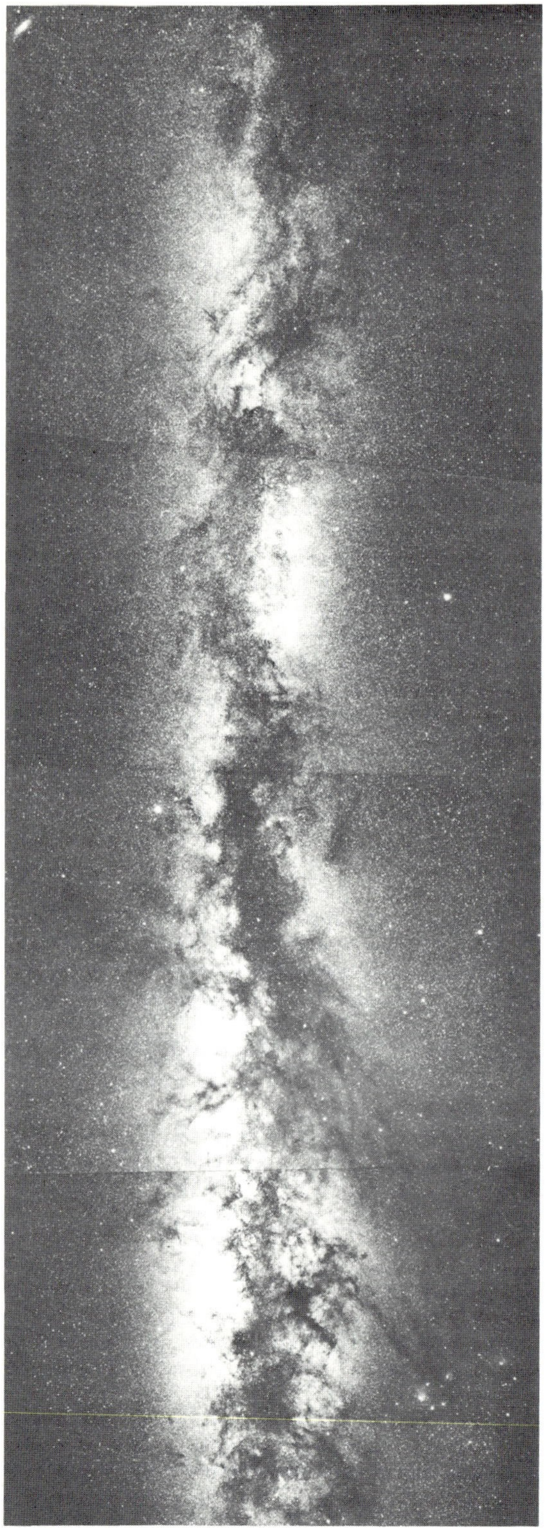

Fig. 2. "Debris" (stars) left in wake of a First Stage mass blown from the original Big Bang sphere. This First Stage mass created our galaxy.

Mount Wilson and Palomar Observatories

THE SOLAR SYSTEM

Our solar system is part of a larger aggregate of stars called a galaxy. The familiar name of our particular galaxy is the Milky Way. Our Sun is but one of 100 to 200 million stars that make up this Island Universe. As well as stars, the galaxy contains massive filaments of gas, luminous and dark clouds of dust and other nebulous matter (Fig. 2). The inter-stellar space, itself, is far from empty; it contains energized particles, molecules, and atoms.

As nature abhors a vacuum, so it abhors a static condition. Our Milky Way, and everything in it, is in motion. Stars are not only turning on their axes but they are also revolving around other points within the galaxy; and the galaxy, in turn, is revolving and speeding through space in respect to more distant galactic systems.

Moving at about 150 miles each second, our Sun and its family of planets takes about 200 to 240 million years to travel around the nucleus of the Milky Way. All stars move at different speeds relative to each other. Stars near the center of the galaxy move faster than those farther away. The orbits of stars are continuously shifting due to the ever changing gravitational fields.

Since all stars are moving in their own orbits in and around the galaxy, and though close encounters between stars would be extremely rare, a close passage would not be impossible. In fact, I am convinced that once every 200 million years our Sun and another star come within close proximity – and at varying distances – to each other. One such encounter, some 5 to 10 billion years ago, created the planets.

The Sun at that time had developed an outer fluid layer of newly-created elements of approximately 5,000 miles in depth. This shroud of liquids and gases had been produced by the Sun over a period of a billion years. It was at about this moment in the Sun's history that from deep space another star approached along a path parallel to that of the Sun. The star, which I will call Mazzaroth, was about ten times the diameter of the Sun and was traveling approximately 1,054,000 miles per hour faster than the Sun. This star passed within a distance of 10 million miles of the Sun, the first of eleven subsequent encounters.

As the two stars passed each other in space, the attraction between them pulled great globules of matter from the Sun's outer surface (Fig. 3). Although the immense gravitational force of the star drew the many liquid-gaseous masses toward it, the radiation pressure from the star prevented their complete capture. Therefore, the effectiveness of gravity attraction was somewhat neutralized by radiation pressure from the star at a critical distance from it, causing all materials following it to slow down and be gathered up by the larger groups of extracted material, or the Sun (since the passing star had a far greater gravitational value, no materials could leave its surface during the encounter).

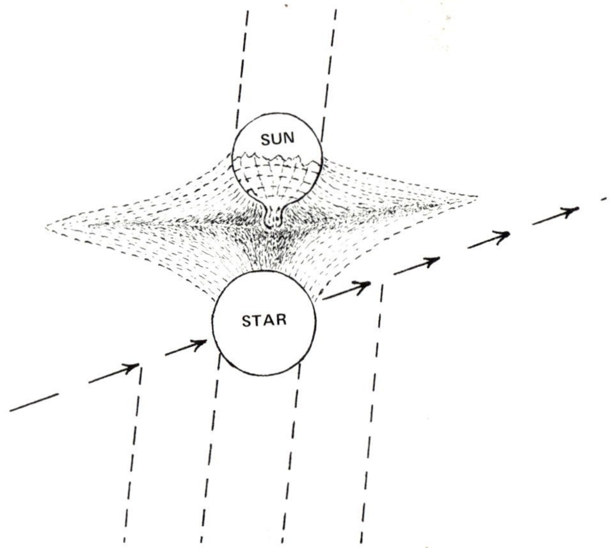

Fig. 3. Depicts forces at play during capture of planetary materials by passing star.

As the passing star moved on into space, the groups of gaseous and liquid materials circled the Sun and in time cooled, condensed and became the planets (Figs. 3, 4).

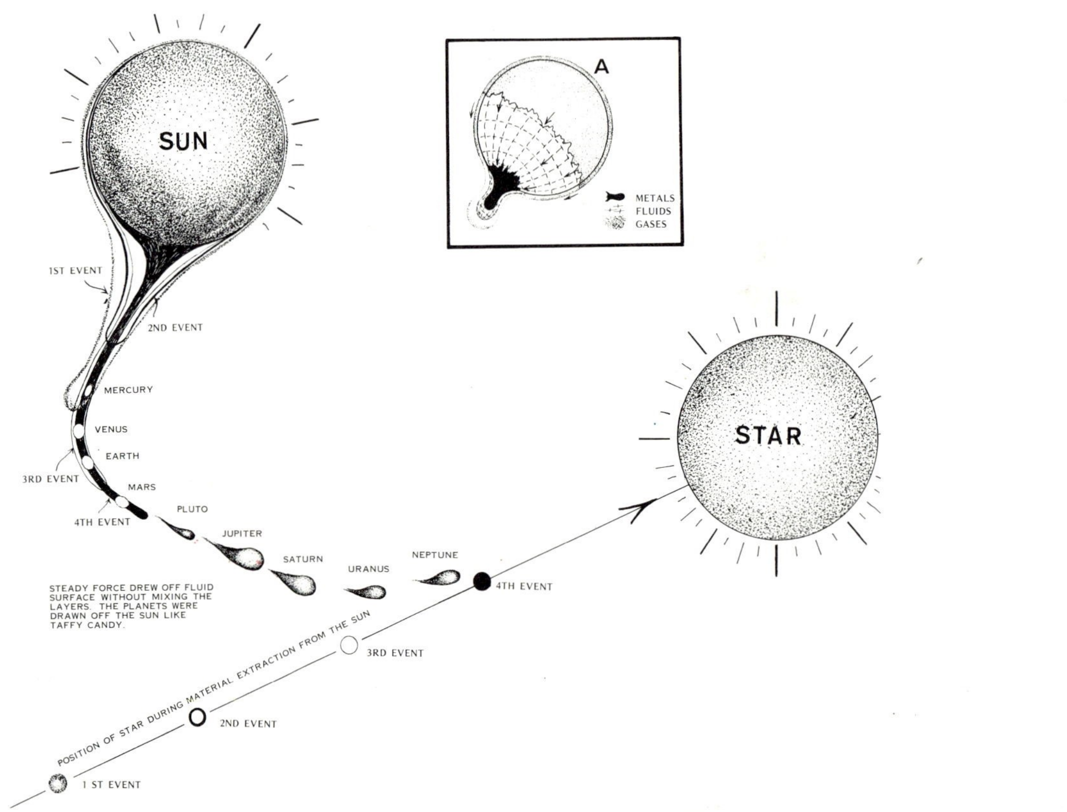

Fig. 4. Balance of shroud "A" reduced in thickness by 50% when the elementsphere moved and leveled over the surface of the Sun after extraction of the planets.

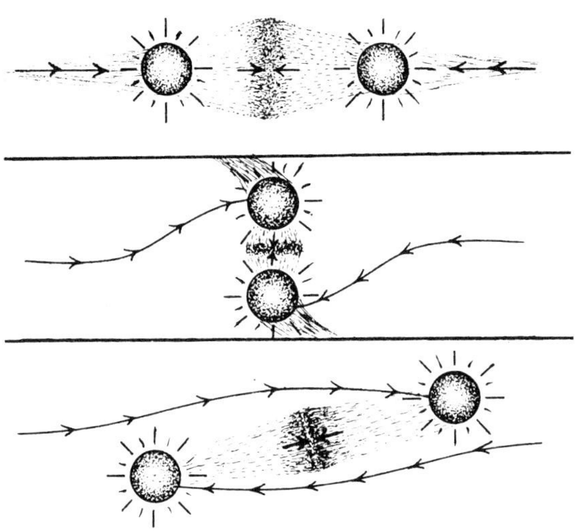

Fig. 5. Radiation forces of two radiating bodies prohibit collision.

In order to fully understand how such an encounter between two stars could create planetary bodies one must first understand how stars are actually structured.

The hypothesis currently propounded by solar scientists is that the Sun and other stars are composed of hydrogen and helium under great pressures, and that nuclear fusion keeps the stars burning. In other words, the Sun is believed to be a great ball of dense hot gas in which neutrons in the nuclei of the hydrogen atoms are constantly colliding with each other and fusing to become nuclei of helium, giving off great bursts of energy in the process. This resultant energy radiates away from the sun (or star) in the form of heat and light.

There are a number of reasons why this hypothesis is not valid, though it is almost universally accepted by most authorities – without question. It is a basic doctrine of solar scientists upon which they build all their subsequent hypotheses and conclusions pertaining to the solar system and other stars. Not one has seriously questioned the validity of this vital and basic premise.

Suffice it to say, many fundamental concepts of physics are overlooked and many basic laws are broken by the accepted hypothesis. For instance, all solar scientists readily admit the paradox of the Sun's temperature. The surface of the Sun is about 11,000 degrees Fahrenheit, the outer gaseous corona is said to be 5,500,000 degrees F, and the interior 28,000,000 degrees F. The Sun is believed to be very hot inside, becoming quite cool on its surface, then becoming very hot again at the outer reaches of its corona. Astrophysicists call this an "interesting paradox." They admit they do not understand how the Sun can go from hot, to cold, to hot.

These temperature changes – from 11,000 F at the Sun's surface to 28,000,000°F at its center – are accepted as scientific dogma without question. This conclusion is pure hypothetical conjecture arrived at to substantiate the hydrogen-helium conversion process.

If the Sun is really a great ball of hydrogen-helium gas as is proposed, then the atoms or parts of atoms making up the Sun must be free to move at tremendous speeds (about 1,000,000 km/sec) in order to produce such high temperatures – three times the speed of light! Yet, how can the particles within such a dense sphere as the Sun (11 times the density of lead) move at all, let alone at such tremendous speeds? And the big question is – what initiates the atoms or parts of atoms into such high speeds in the first place?

It has been stated that the extreme temperatures within the Sun would cause electrons to be stripped from all atoms. Hydrogen and helium would no longer be true atoms but only the bare proton and neutron systems of hydrogen and helium (one of the basic nuclei or components of all atoms).

But – if electrons were stripped from the atoms of hydrogen and helium leaving only their nuclei, these nuclei would not "take up less room" i.e. pack tightly together, as is proposed in order to account for the high interior density of the Sun, but, instead, would repel each other, thus causing the Sun to expand into nothingness in the vacuum of space – in other words, it would blow up.

If the interior of the Sun actually is a very dense material under great pressure and temperature, how could this material still be a gas? How could any particle move under such pressures, and if so dense, where would the space be in which these particles could so freely move? And it is a basic physical law that without movement high temperatures cannot generate.

Furthermore, within such a dense mass, nothing exists that could initiate enough movement to create even one collision between particles, let alone the countless high speed collisions required to produce the proposed high temperatures said to exist in the Sun.

Also, due to the nature of protons, neutrons, mesons, muons and many other sub-systems, all of which are inherently endowed with a limit of energy (are fully saturated), all systems are in equilibrium and at neutral. They cannot emit or receive energy when locked together tightly.

In addition, there is the problem of the many other elements produced by the Sun. If these elements are produced in the interior, a build-up in quantity would progressively inhibit the fusion process until it reached zero.

The ratio of chance of elemental conversion would be very low since a specific number of "hits" *continuously* maintained by *specific elements* would be necessary. A hit and miss collision process could not maintain a constantly regulated surface temperature. Furthermore, this thermonuclear method does not account for the sunspots nor for the regular cycles of them.

Lastly, extremes of temperature are erroneous – the maximum thermal attainment ends at the speed of light, since for every action there must be an equal reaction. That is, for every degree rise in temperature in any mass a specific amount of energy must be supplied. These actions are governed by the *space* in which energy may act upon matter in order to raise its maximum temperature. (Fig. 6).

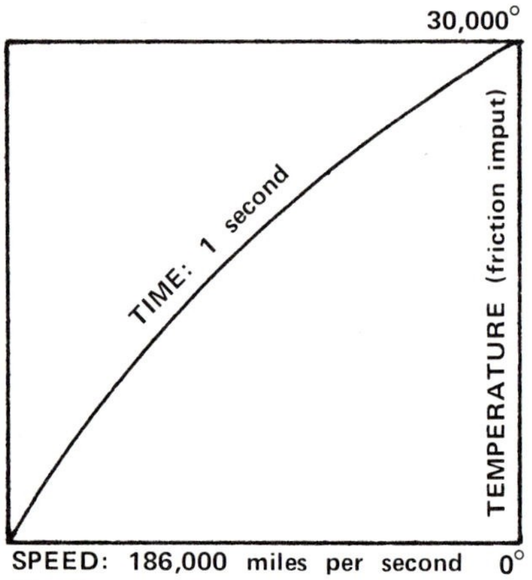

Fig. 6.

The temperature attainable from absolute zero (when all systems of atoms are at rest) to maximum is approximately 30,000 degrees F. This being the point at which the expansive rate of all parts of matter achieves the speed of light, which is also the maximum speed energy may impact to generate thermal radiance. Above 9000 degrees F all elements have gone into a gaseous state. At 20,000 degrees F, molecular units have separated and the atoms are greatly expanded.

The space, due to distention (expansion) between the parts of matter absorbs and transports much of the energy input. As distention of matter increases with energy input, so increases the speed of matter to the point that the volume (total quantum) of energy reaches a balance with the speed of matter. In addition, energy is ineffective and lost in the space vacated by matter distention – thus the limit of attainable temperature is reached. The best man-made temperature detector of any reliability can measure little more than 6000°F with any accuracy. Any equipment devised to measure temperature above this is generally based upon theoretical conclusions from some photoelectric effect, or other such devices.

Extremely high concentrations of brilliant light from a central source (star or hydrogen bomb) emanating along with extremely high amounts of energy, are misleading. Light is not heat, only evidence thereof. A body or mass at 30,000°F would be at its extreme of brilliance. The brightest stars in the universe cannot be hotter than 30,000°F.

In addition, the activity of millions of spicules constantly in action on the surface of the Sun belies the hydrogen-helium conversion process, for these are mainly calcium and silica materials shooting up at high speeds, partially going into a gas or retracting – all within a few minutes. This, of course, is the first liquid matter below the gaseous outer envelope that overlies all bodies in the universe, stars as well as planets. (Below the Earth's organic crust begins the granite layer which is composed mainly of calcium and silicas.)

Could it be that the accepted theory of the Sun's structure and material makeup is wrong – that perhaps its interior is not hot at all but actually quite cold? How can the Sun be a great ball of hot gas? It is too dense and it is too stable. Its outermost atmosphere may be gaseous but the main body of the Sun must be something else. So let us consider a Sun that is structured in a manner that is more in agreement with the natural laws of the physical universe.*

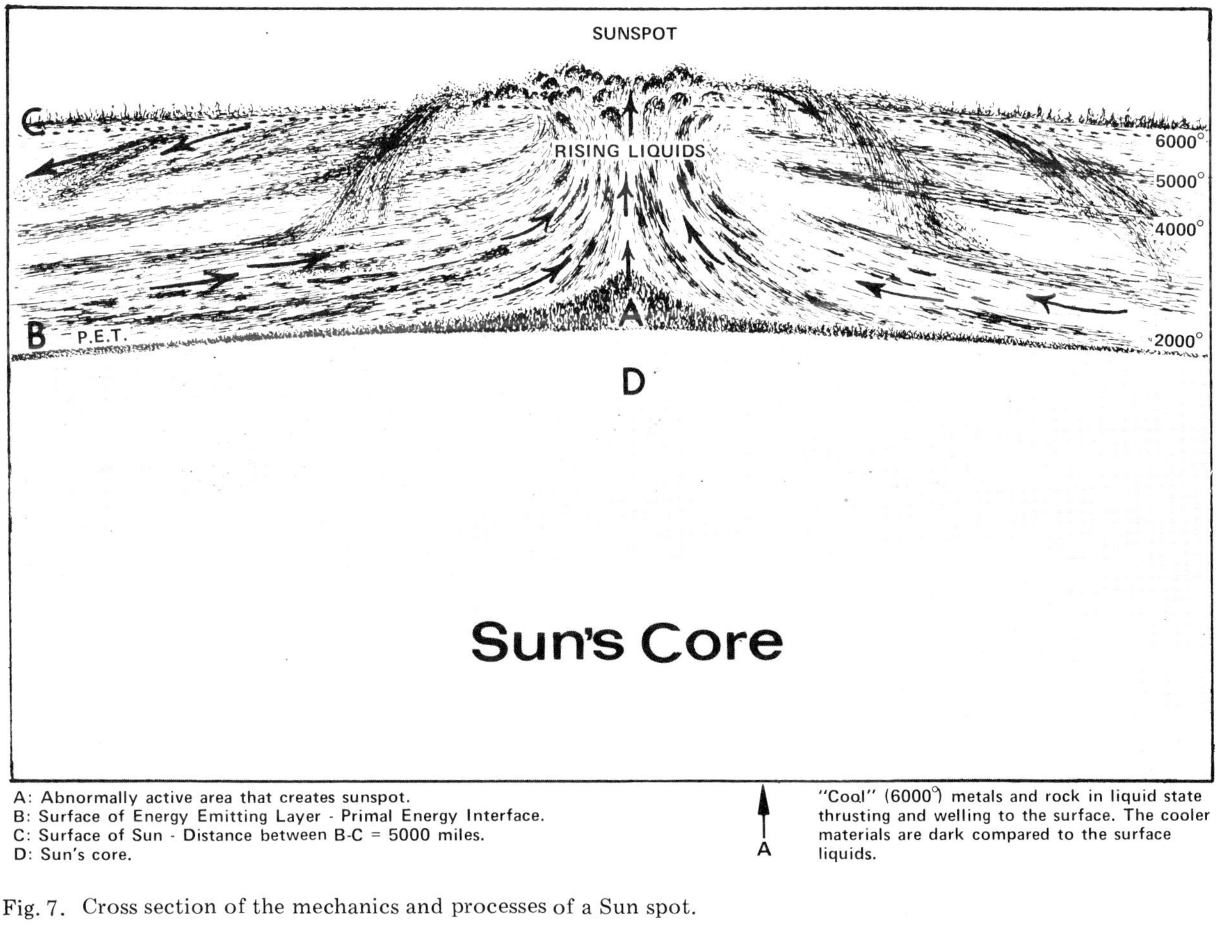

Fig. 7. Cross section of the mechanics and processes of a Sun spot.

* Covered in detail in THE LIVING SUN and THE THEORETICAL STRUCTURE AND CYCLE MECHANICS OF STARS by Lewis A. Manson (in preparation).

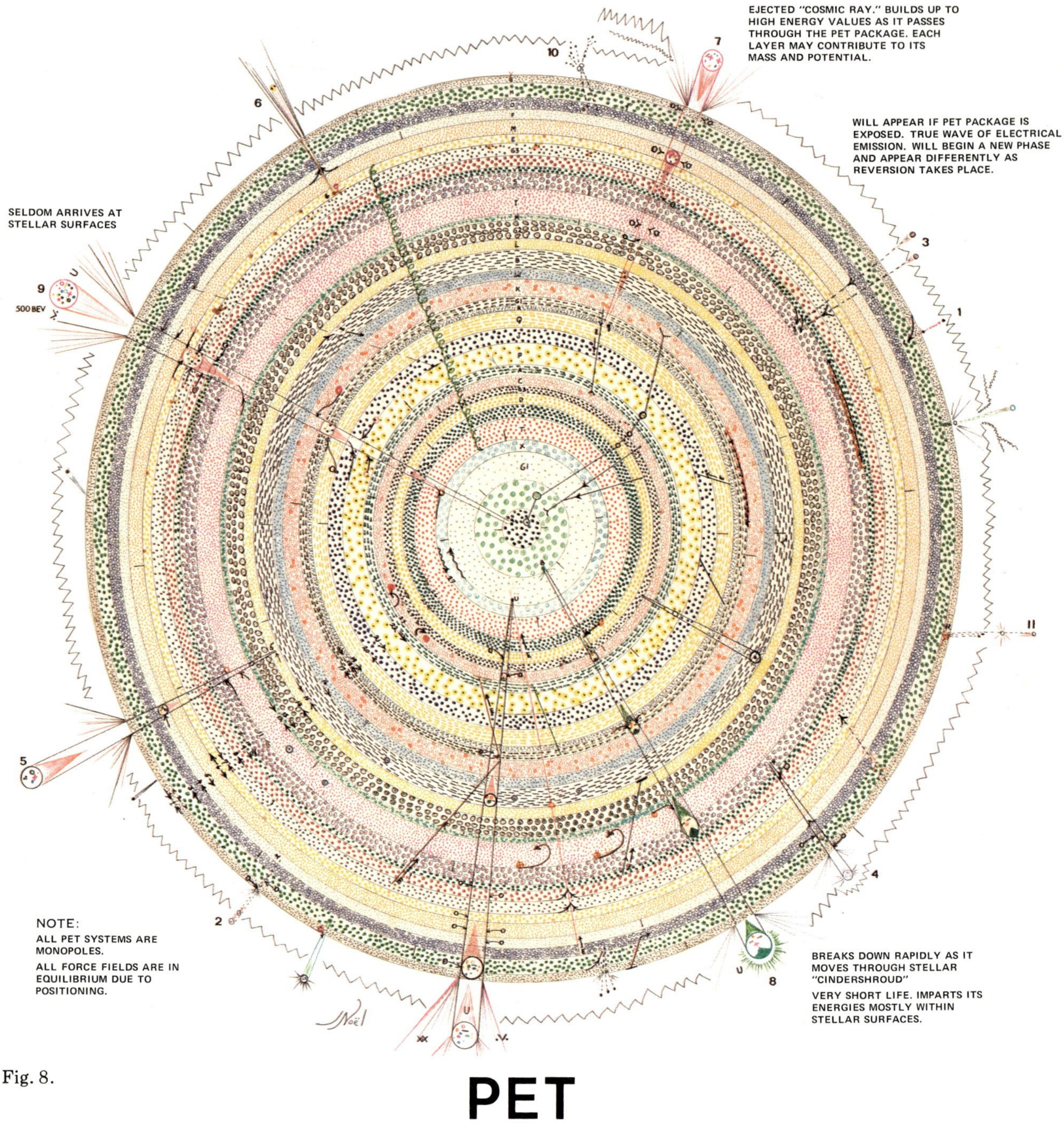

Fig. 8.

PET

THE PRIMAL ENERGY TRANSVERTER*

The above drawing is a cross-section of the Primal Energy Transverter (PET),a spherical unit about the size of an atom. PETs are composed, totally, of energy functioning in a great variety of unique circuits. Their emissions of energy can only take place when they become exposed to non-emitting or lesser emitting areas. If they are surrounded by other like systems they cannot begin energy emissions. On the Sun only at the primal energy "interface" with the cindershroud (elementsphere) do these emissions take place. By the time the energy of a PET has been expended, it has become a "cinder" of the energy package and, therefore, an element.

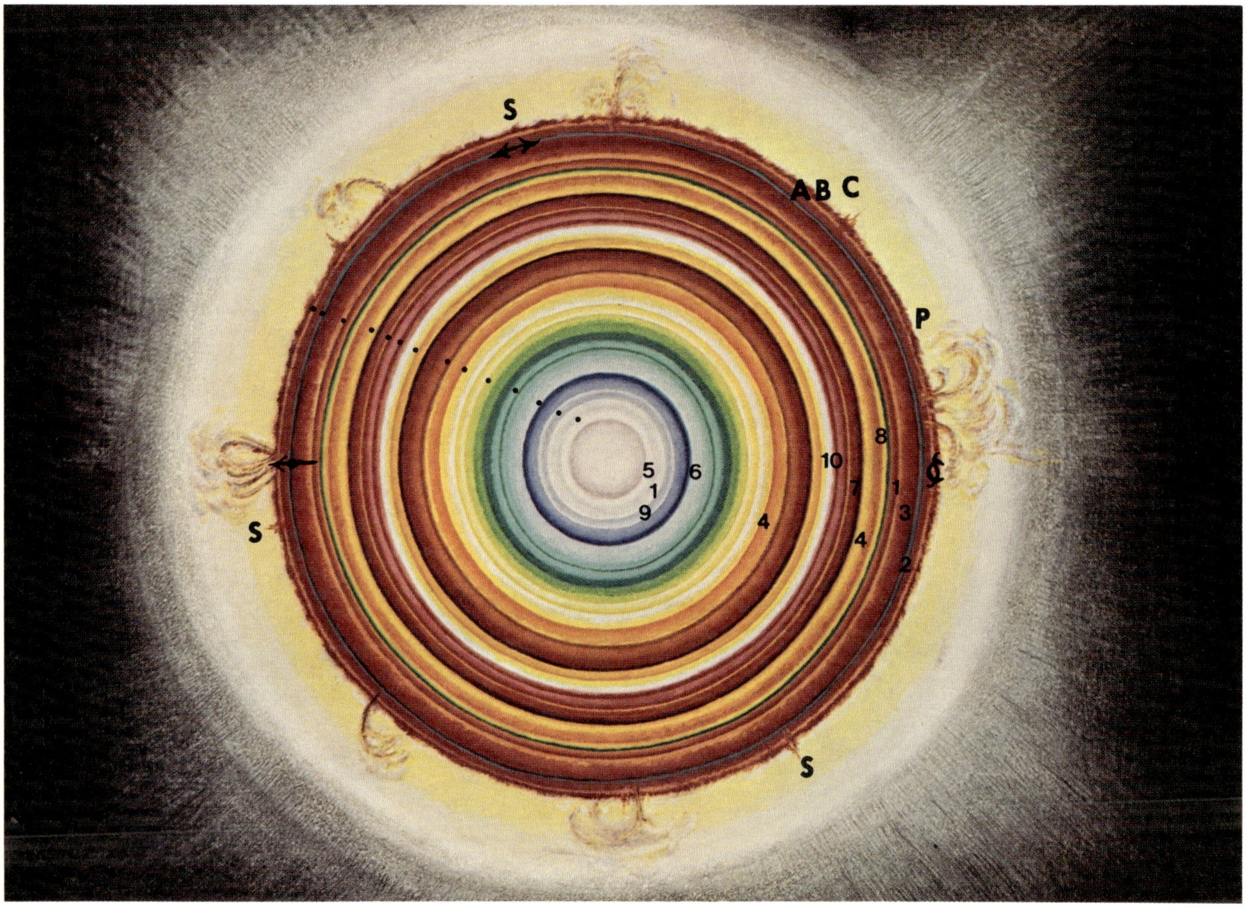

Fig. 9. A CROSS SECTION OF THE SUN'S STRUCTURE

The above illustration is only representative of hundreds of layers (pre-scheduled cycles).

KEY TO SYMBOLS

- P Photosphere
- A Heavy liquid elements } Elementsphere (energy cinder shroud) approximately 5,000 miles deep
- B Heavy gaseous elements }
- C Light gaseous elements: Chromosphere
- D Hydrogen and Helium layer: Corona. Origin of most "starlight" due to concentration of energy emissions which cause the layer to glow to extreme brilliance.
- S Spicules: 2,000,000 to 5,000,000 of these are evident over the total surface at all times. These are the light liquid elements boiling. Gases cause same effects as boiling water.*
- ↔ Surface where emission of all energy occurs (currently) :Primal Energy Interface with the Elementsphere.
- 1. Current emitting layer.
- 2. Layer of Novae potential.
- 3. Creates large volumes of metals, less light elements; also high magnetic activity.
- 4. Unknown, unless this would be the minimum mass size of radiators.
- 5. Heavy creator of light elements. Energy emissions not so intense.
- 6. Pre-Novae layer: extremely intense emissions of energy—a thick layer but of short cycle. Low metal but high gas creator.
- 7. Hard beta, gamma and cosmic ray emitter. Less matter created.
- 8. Creates full range of elements in large volumes. Heavy energy emissions. Creates effects quite similar to the Cretaceous Period on the Earth.
- 9. Great radio emissions. Creates extreme heat at surface and drives off surface debris steadily. Heavily ionizes atmosphere on Earth.
- 10. Period of relative quiet. Soft radiations. Heavy creation of gaseous elements.
- ℄ Elementsphere deeper at the equator due to centrifugal force. Few sunspots visible in this area. This is also the dividing point of
- -•→ direction in which surface liquids circulate (currents of the elementsphere).
- • Transition points where energies of ending cycle blend with beginning of new cycle. Various phenomenon occur here. Some surface clean-off possible. Age of each cycle varies from 5 to 10 seconds to 300 million years duration.

Each layer of the Earth's organic crust** represents a distinct and separate cycle of the Sun. Each cycle has produced a different form of life, set of circumstances, grade of material deposition. Each cycle represents a transition from one layer to a distinctly different layer on the Sun. This information is based upon the kinds of materials (basic) that arrived during each geological era.

An electrical force of unique characteristics goes into action periodically and realigns the materials just below the energy emitting area (2) putting the powerful materials in "make-ready" condition. The materials, that have been "activated" by this force, go into action as the emitting surface gradually creeps downward into this deeper area of the layer. Without the re-alignment, the emission would not occur. Evidence of the realigning force in action is in the 11 year cycles of "sunspots."

* They are composed of calcium, silica, sodium and lesser amounts of other low specific gravity elements.
** This applies to the other planets as well.

★See THE LIVING SUN (in preparation)

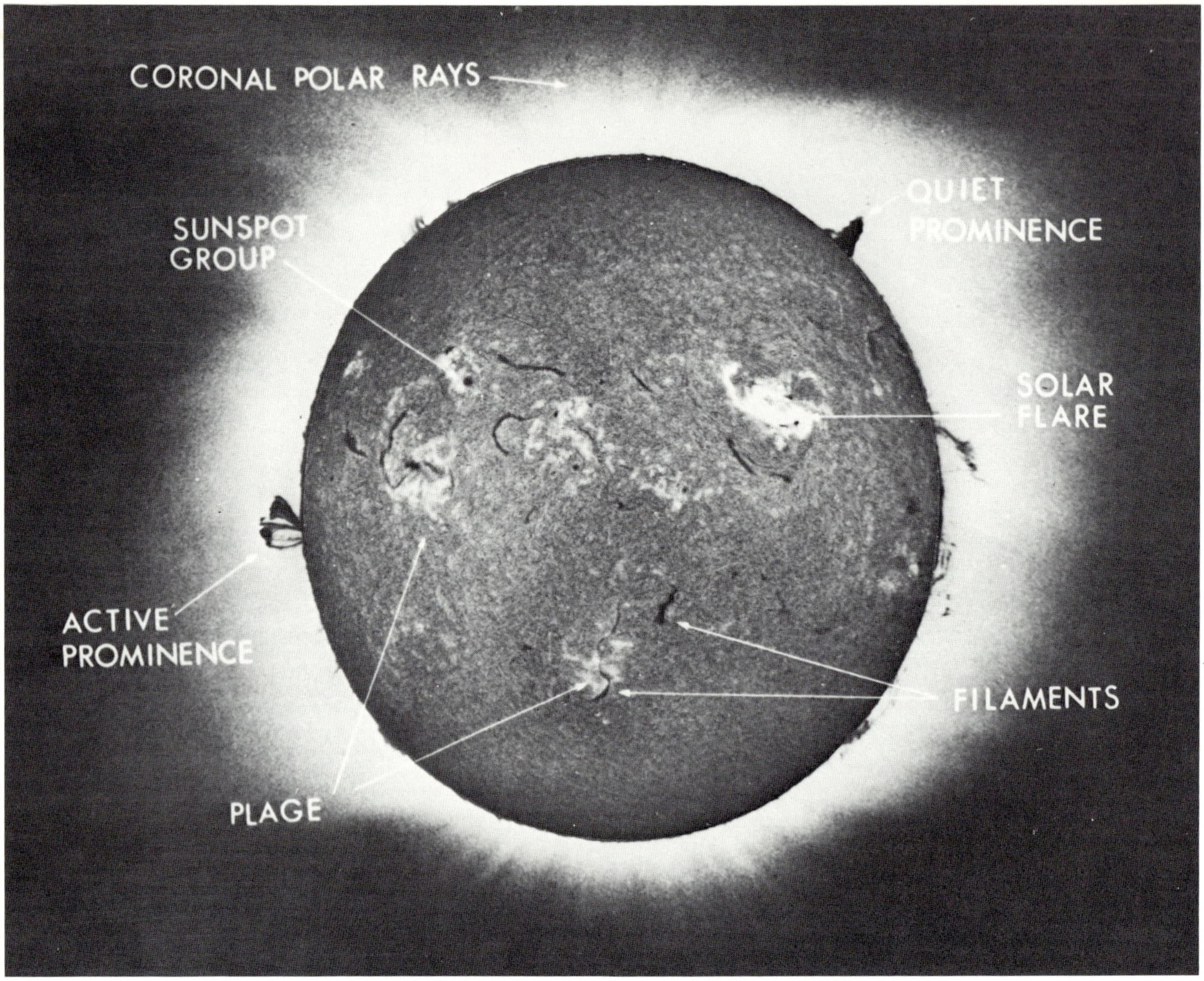

Fig. 10. Special composite photograph of the Sun. The solar disk was photographed in hydrogen light on July 16,1959 and the corona on July 29,1963. The prominences were taken at the Mt. Wilson observatory.

Lockheed Solar Observatory
Lockheed Aircraft Corporation

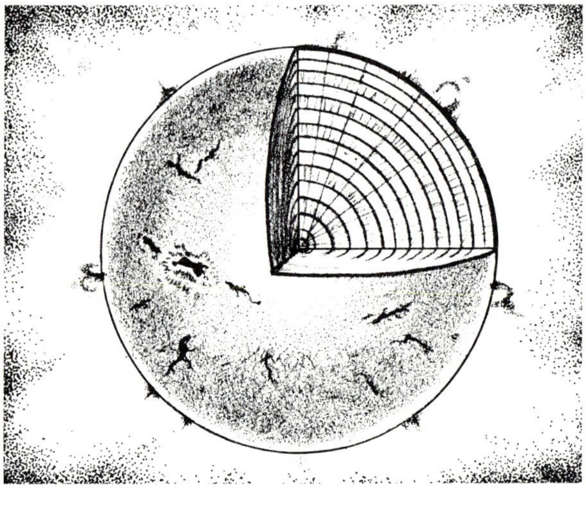

Fig. 11.

Fig. 12. Stages of activity of the Sun

Let us examine a Sun that is composed of many intricate layers of densely-packed "energy systems." These layers are force shells, in between which are strata of potential matter-producing energy systems. These systems, as a result of their functioning, create particles (atoms) of matter. Each layer within the Sun produces an entirely different set of circumstances, activity, affect and effect. Each layer is a cycle "in waiting," replacing the outer layer after it dissipates its energy. Some cycles end quietly with a new one beginning almost without notice. On rare occasions other cycles begin and end with tremendous nuclear explosions (novae) (Figs. 16-A and 16-B) which automatically disperse the heavy accumulation of elemental debris from the surface of the Sun, or other star, (all stars experience these cycles). This debris is the natural result, or by-product, of long cycles of energy emission.

Each layer, because of its position and composition, has a specific function. Only the outer layer of the Sun is violently active. (See Fig. 9 which is a cross section of the Sun showing its structure.) This outer layer, which I have named the "cindershroud" or the "elementsphere," was built up over billions of years. It is a layer of liquids, 5000 miles deep, made up of all the 106+ "elements." This layer seethes with activity and shrouds the high energy emissions coming from the Sun's interior. This shroud of liquid elements is a natural function of the Sun and governs its energy emissions and acts as a regulator in several ways. Some of the most important ones are:

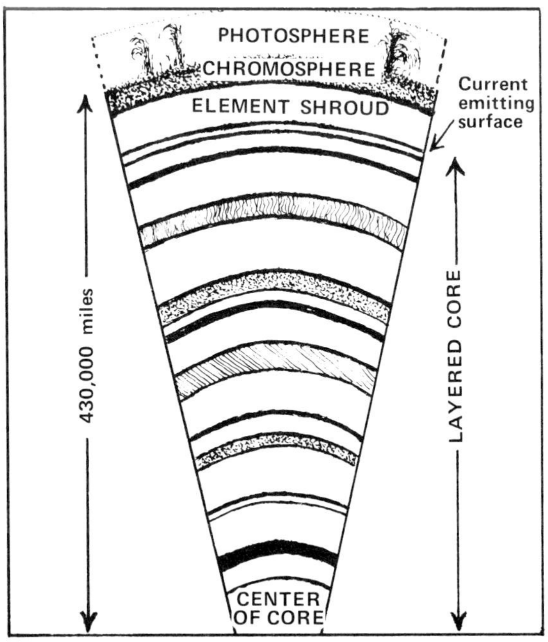

Fig. 13. Cross section of the Sun showing its layered structure.

1. The thick layer of elements assure an even radiation from the surface and the constantly maintained high temperatures create an even mixture of the energies in the surface layer.
2. The thick layer of elements allow the creation and propagation of light.
3. The thick layer of elements prevent a nova from occurring prematurely. The blanket of liquid elements fills all cracks and cavities in the main energy emitting surface thus allowing each cycle to complete its full life without interruption. Any cavity created in any of the energy layers to a depth of the next layer would instantly cause an explosion. This explosion would immediately strip off the balance of the entire elementsphere; it may even penetrate more than one layer.

Below the cindershroud is a zone which I have named the "Primal Energy Interface." This is the area where the emissions of the primal energy transverters (PET's) move into the cindershroud (see the cross-section of a sunspot, page 25 and the cross-sectional drawing of the Sun on page 27).

From the energy emitting surface (Primal Energy Interface) downward all the way to the center of the Sun are hundreds of layers composed of primal energy transverters. Each transverter is about the size of an atom. Through eons of energy emissions from their enormous energy potential, these tiny systems have converted to elements making up the "cindershroud" (elementsphere).

Each PET system is endowed with specific kinds of internal circuiting arrangements of energies so that they create a specific kind of element. These tiny systems are arranged in the layers of the Sun with each layer designed to be a specific cycle of the Sun's activity which produces, by its energy emissions, varying types of radiation emanations which create specific kinds of environments on the Earth and other planets. Each cycle may last from 5 seconds up to 300 million years. For example, the period which built the rocks of the Cretaceous Period lasted for over 200 million years.

The PET systems are the answer to the energy displayed by stars in nova stage, since energy from the PET systems unlike the proposed helium-hydrogen reaction in stars, does not originate at great depths, but only at the base of the cindershroud and instantly strips away this cover, exposing the radiating layer.

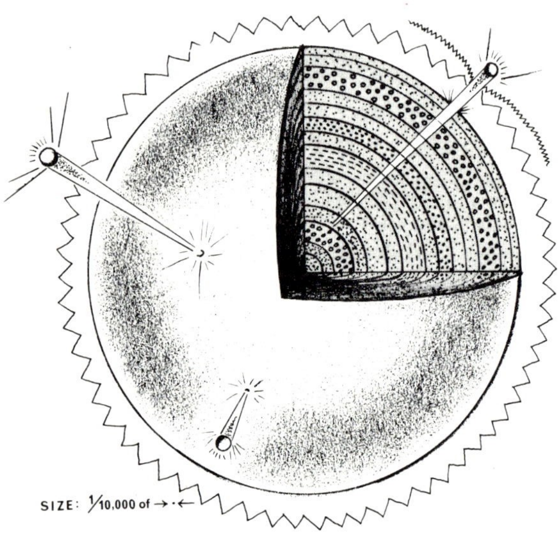

Fig. 14. The Primal Energy Transverter (PET) functions on the same principle as does the Sun. The only difference is the kind of function.

Nova cycles are rare in the life of a star, because the spherical design of a star serves to assure perfect emission of energies away from the surface without interfering with any of the emissions from an adjacent surface area. The spherical design also allows equal dispersion of each internal layer.

Over the elementsphere lie the gases of the Sun. These gases are: hydrogen, helium, neon, oxygen, xenon, calcium, silicon, and small amounts of many of the elements of low atomic weight. These gases are always in great turbulence due to the extreme heat generated by the energy emissions that flow through the elementsphere. These energies constantly maintain high friction impact with the atoms of the gases of the photosphere and chromosphere, thereby maintaining the high temperatures of the Sun's outer surface.

The function of the PET can only be briefly told here, but must be touched upon since it is the "key" to how the Sun and other stars function and leads to clearer understanding how the beautiful and complex universe is built. This extremely brief outline is taken from the more detailed book on the subject.*

* See "Primal Energy Transverter" by Lewis A. Manson (Copyright - 1966)

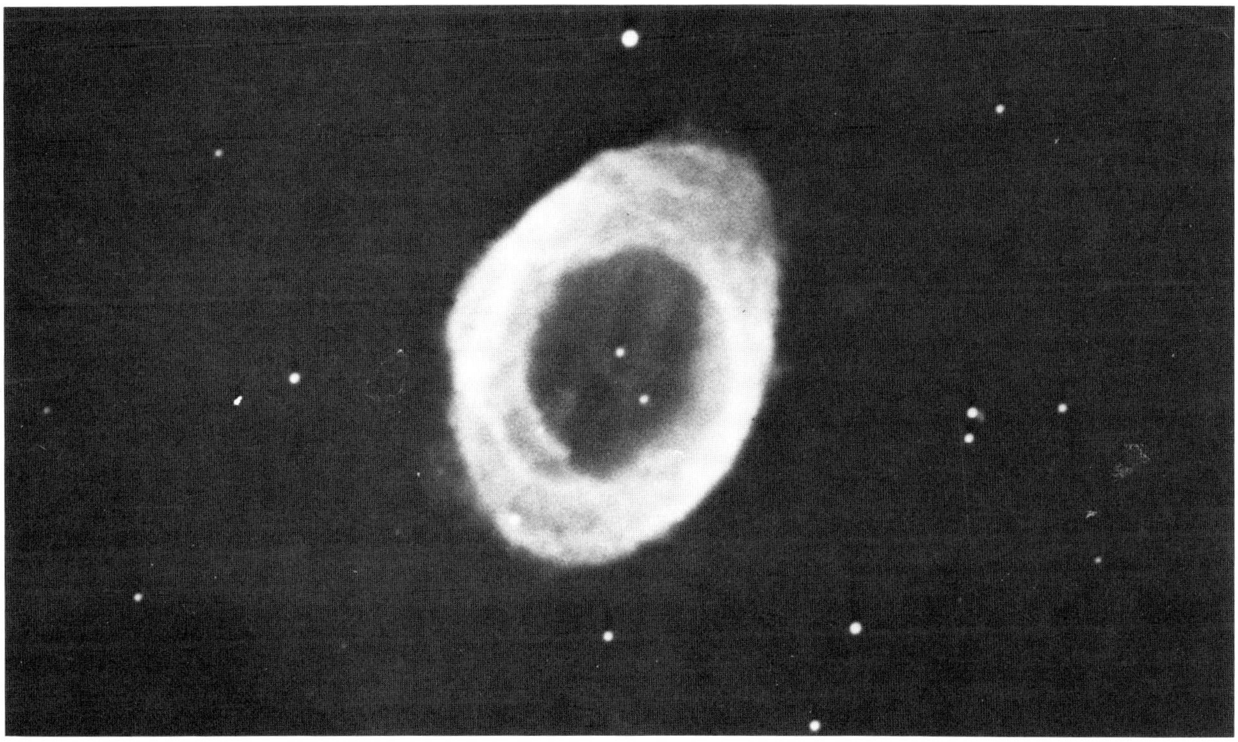

Fig. 15-A. The Ring Nebula. A post-nova state in which the strength of the nova being lower than normal expulsed only the gases from the "shroud" of the star. **Mount Wilson and Palomar Observatories**

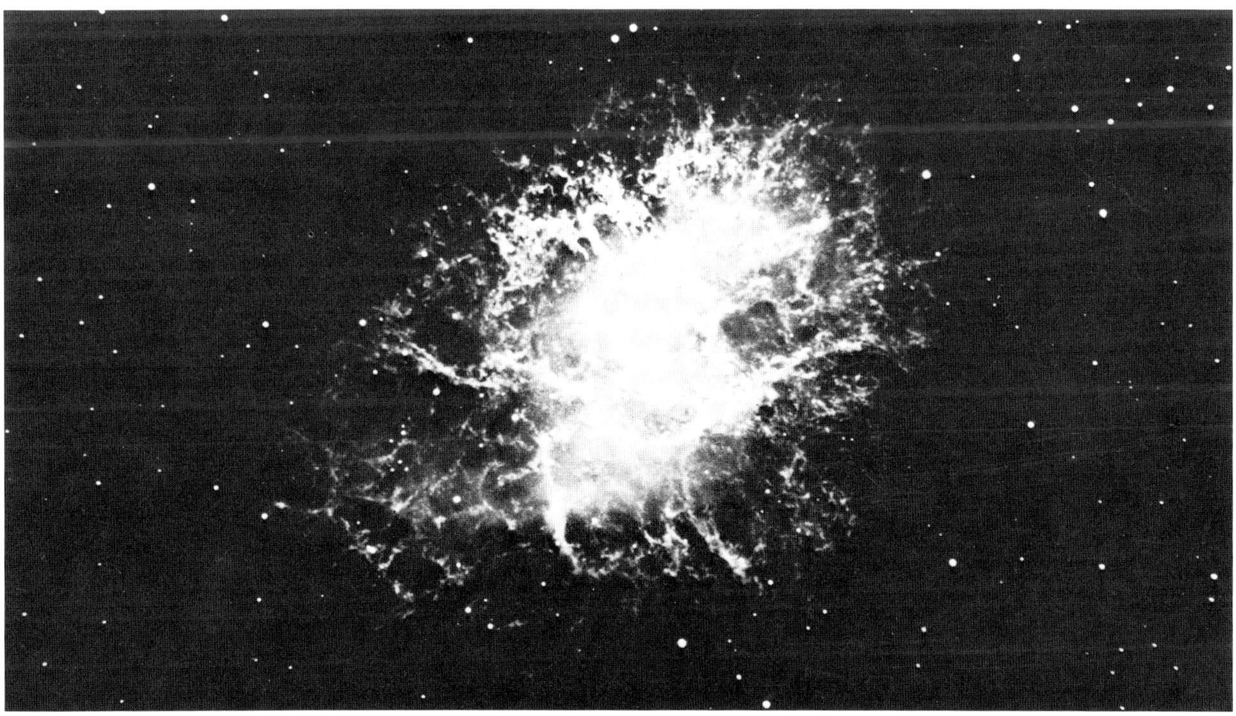

Fig. 15-B. The Crab Nebula. In 1054 AD this star went into a massive nova which completely blew off the total cindershroud (elementsphere) of its surface. The explosion explused enormous volumes, or quantities of gases, fluids, including huge amounts of metals and radioactive debris. Radiation pressure from the intiating star prevents the expulsed materials from returning to its surface. **Mount Wilson and Palomar Observatories**

Within the elementsphere (cindershroud) are areas (layers) of every element (although constantly distorted by fast rising currents from below). The preponderance of the elementsphere is made up of partially degenerated energy materials (extremely radioactive) emitted from the energy emitting layer. These materials have gone into and mixed with the fluids in the elementsphere and make up 90% to 95% of it. The balance is composed of varying ratios of every element.

When material from the elementsphere was pulled away to create the planets, the extremely dense energy material was in a liquid form and immediately sought the central area of each globe of the gaseous and liquid elements to become the core of each newly-created planet. The elements cooled and arranged themselves in the order of their specific gravities into a crust over this highly radioactive core. This is the source of the internal heats of each planet.

Were it not for the "cindershroud" – this deep layer of liquid and gaseous elements, the Sun could not have given birth to the planets. If the Sun actually is (as is proposed) nothing but a great ball of hydrogen-helium gas, the 92 elements which make up each of the planets in varying proportions would not have been in existence on, or in, or even near the Sun at the time of the planets' creation regardless of which theory of creation is preferred.

The enigma of the creation of the planets has always been with us and consequently has produced numerous theories, most of which were wrong simply because the structure of the Sun was based upon the results of nuclear experiments here on Earth and from these experiments details on the structure of the Sun were subsequently developed. But with a new and more logical concept of the true structure of the Sun we can now begin to understand how a passing star could produce planets from the Sun under the most natural of conditions.

Fig. 16-A. A star prior nova stage.

Lick Observatory

Fig. 16-B. Same star in nova stage a few weeks later. The debris had expanded many millions of miles out into deep space.

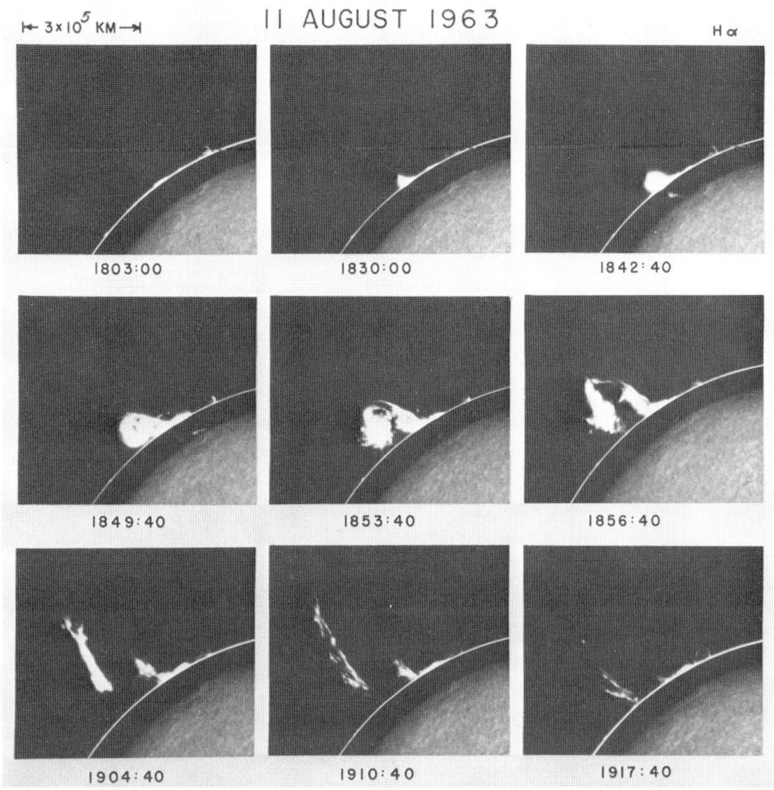

Fig. 17. Massive expulsion of materials in stages. This prominence was created by a short-lived class of sunspot in which the materials shot out from the Sun several hundred thousand miles.

Lockheed Solar Observatory
Lockheed Aircraft Corporation

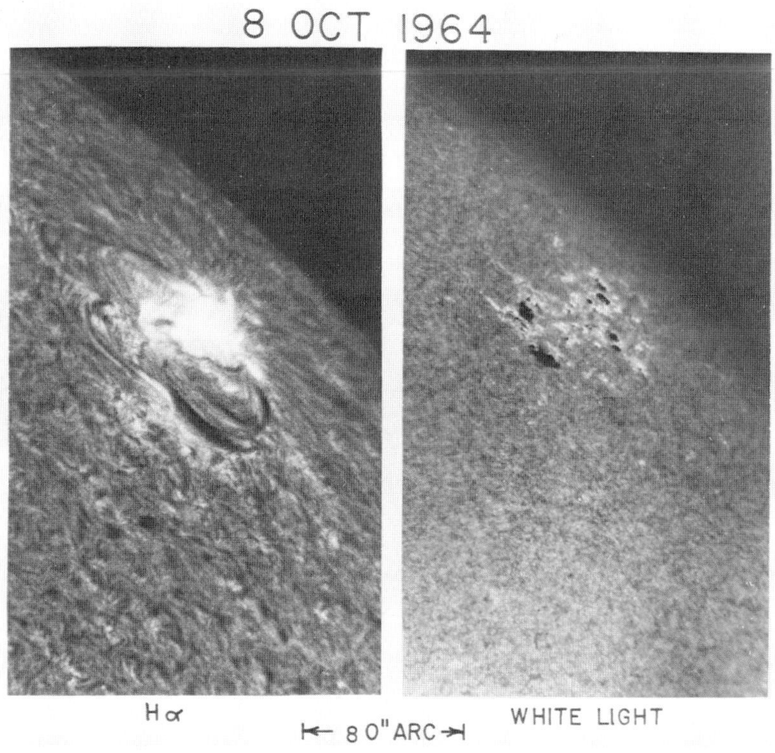

Fig. 18. Top view of same process. Prominences can be composed of calcium plage, silicas or even metals at extremely high temperatures.

Notice the high rings of waves created by the welling up to the surface of cooler fluids.

Lockheed Solar Observatory
Lockheed Aircraft Corporation

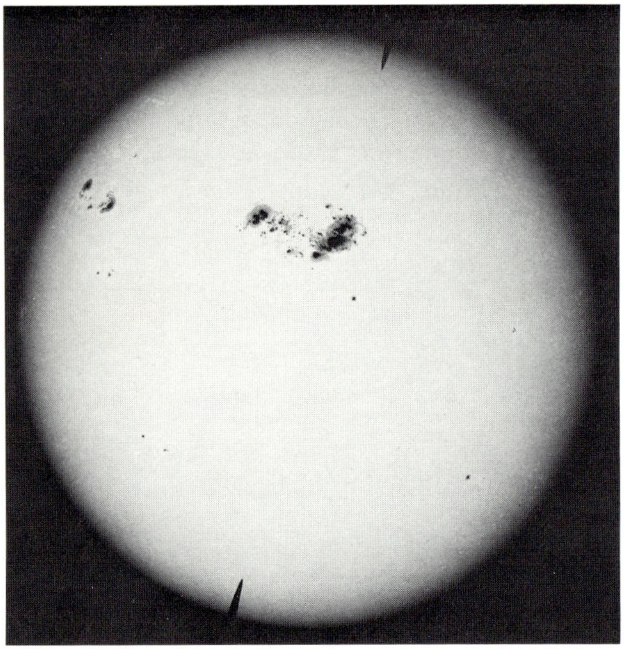

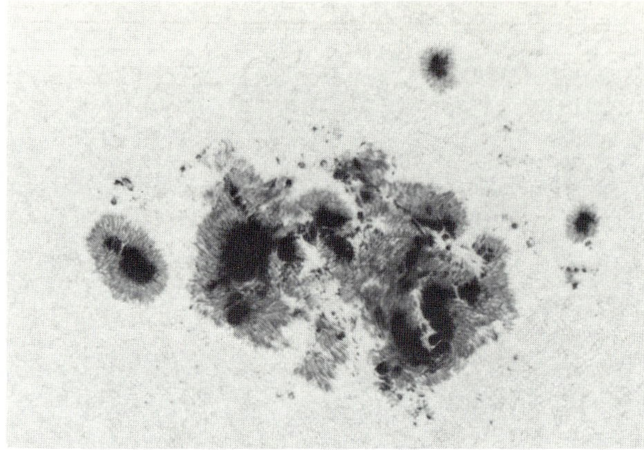

Fig. 20. Looking directly down into a series of sunspots. The black areas are "cool" metals arriving at the surface. The metals originated from about 4000 to 5000 miles down. The grey areas around the umbra are these metals blending in with the surface fluids of calcium and silicas which are very hot (11,000° F.

Fig. 19. Sun with sunspots.

Mount Wilson and Palomar Observatories

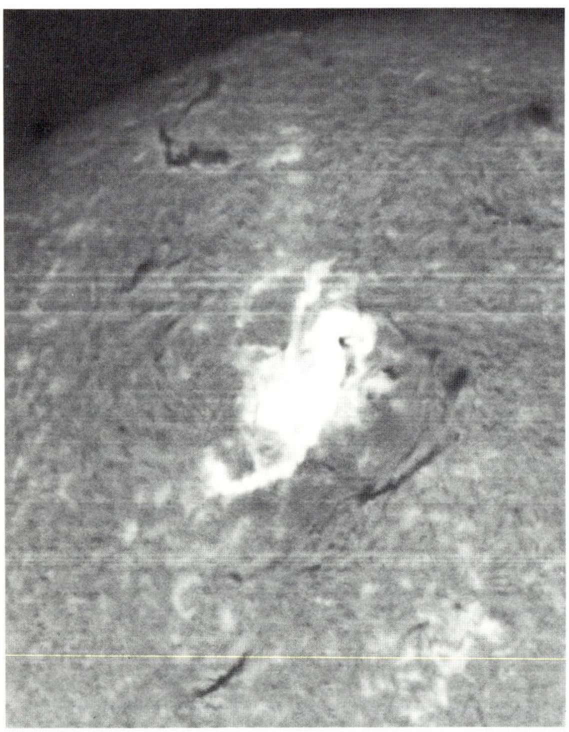

Fig. 21. Primal energy material arriving at the surface and mixing with the surface materials. The brilliance of this flare is due to minor fissioning taking place.

Mount Wilson and Palomar Observatories

Fig. 22. The magnetic lines of force are evident here in the materials flowing away from the edge of the sunspot. These magnetic fields are created by the extremely high energies emanating from the sunspot.

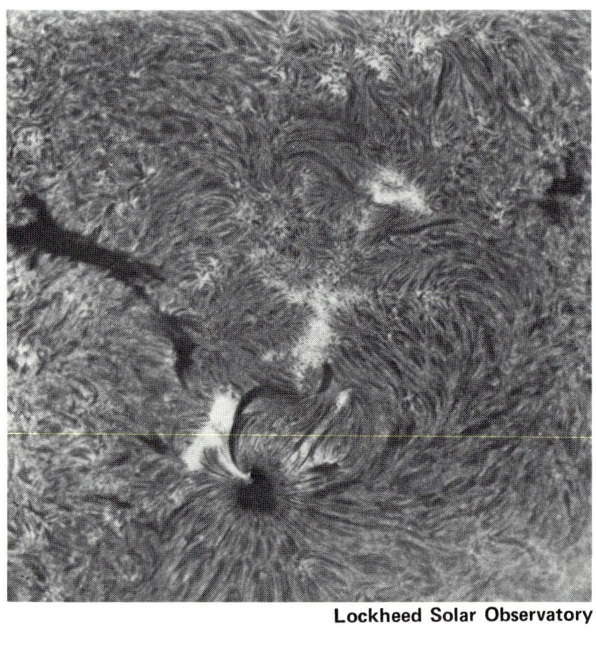

Lockheed Solar Observatory

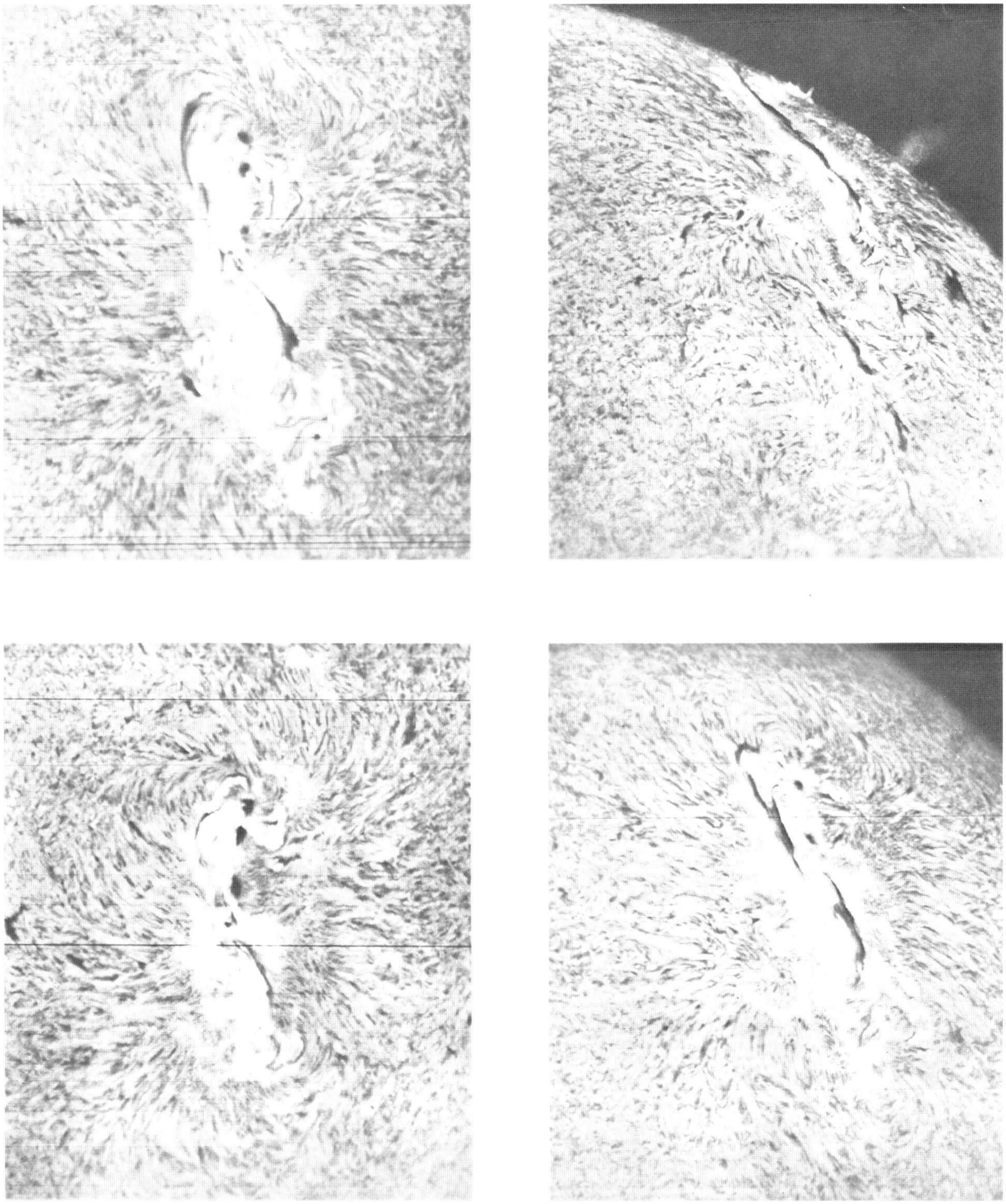

FLARE ACTIVITY ON THE SUN

Fig. 23. Mount Wilson and Palomar observatories

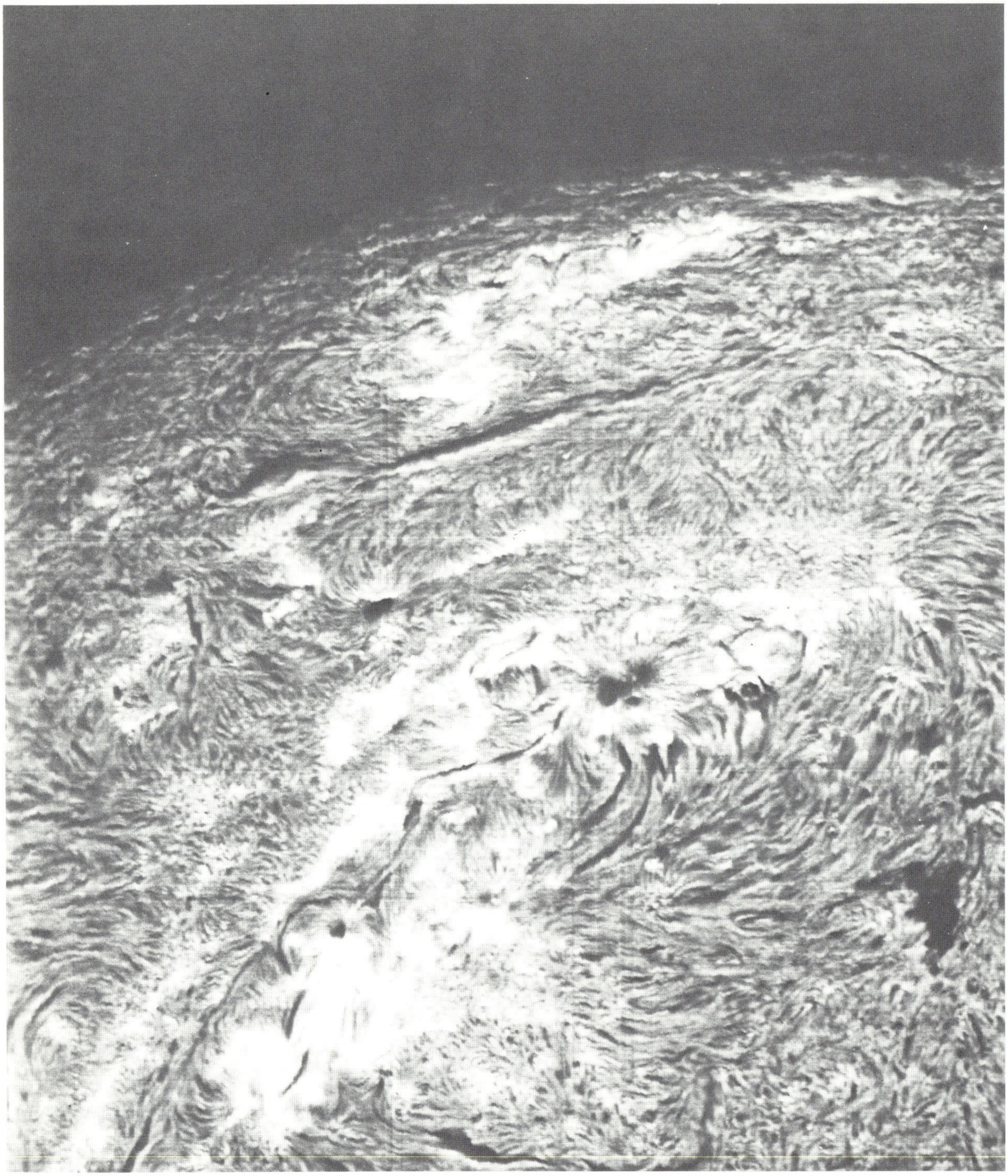

THE SUN'S TURBULENT SURFACE

Fig. 24. Mount Wilson and Palomar Observatories

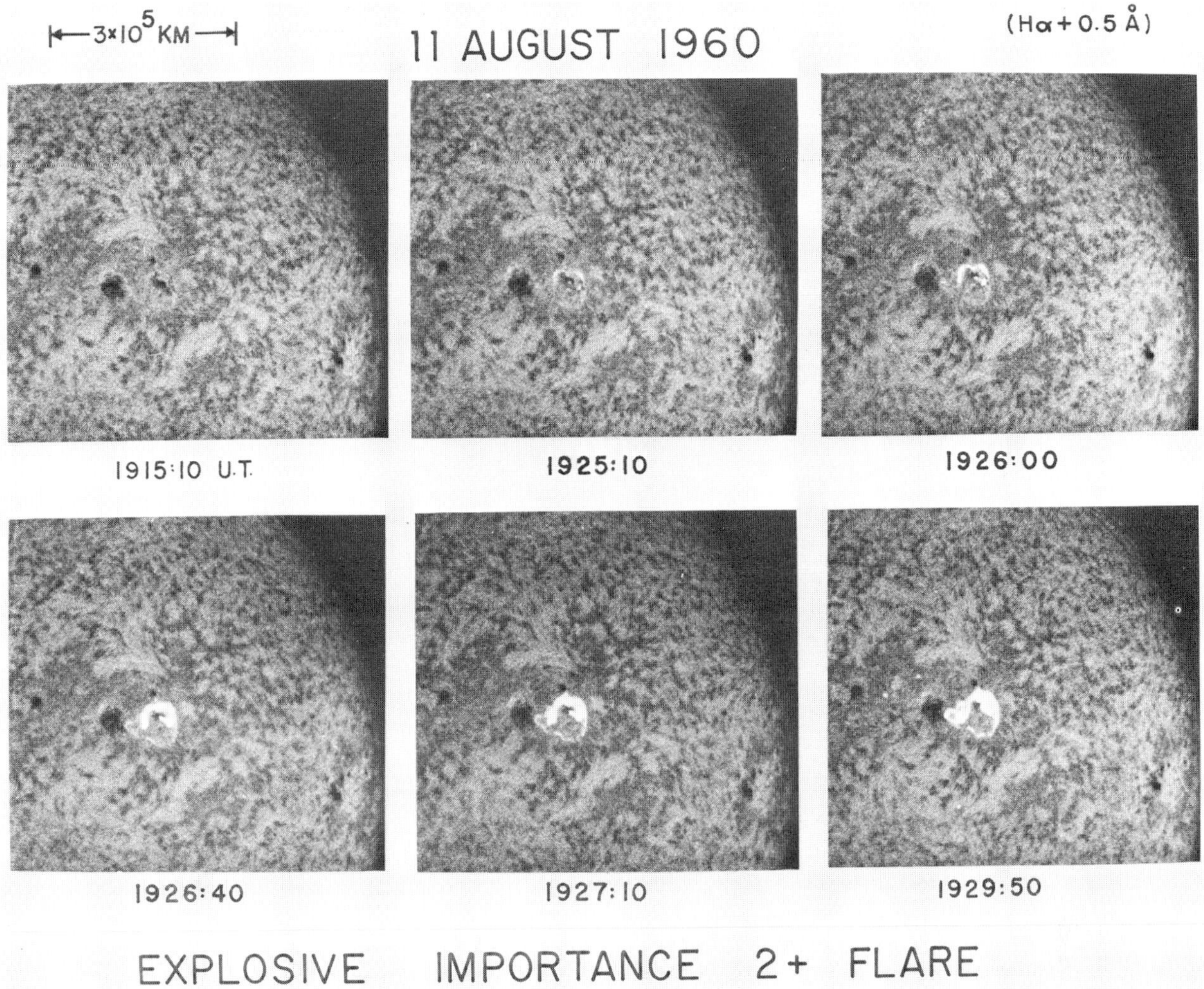

Fig. 25. Lockheed Aircraft Corporation

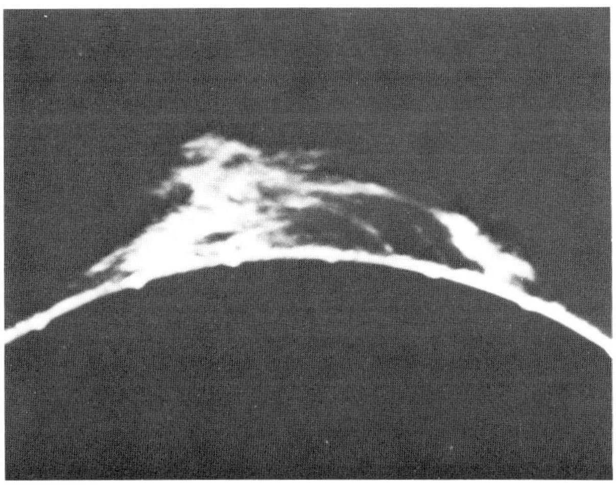

Fig. 26. Solar prominence 132,000 miles high

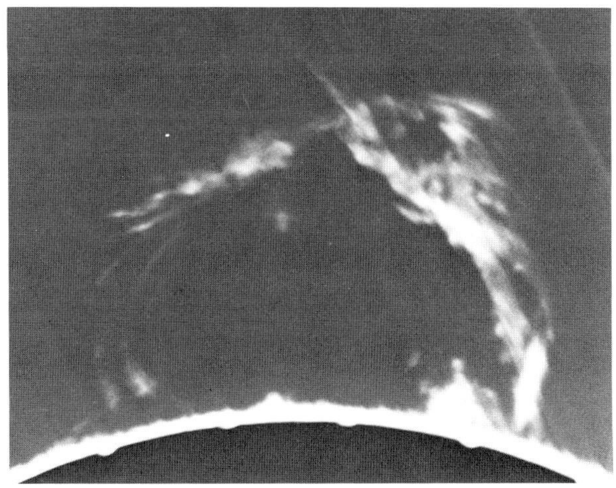

Fig. 27. Solar prominence 205,000 miles high

Mount Wilson and Palomar Observatories

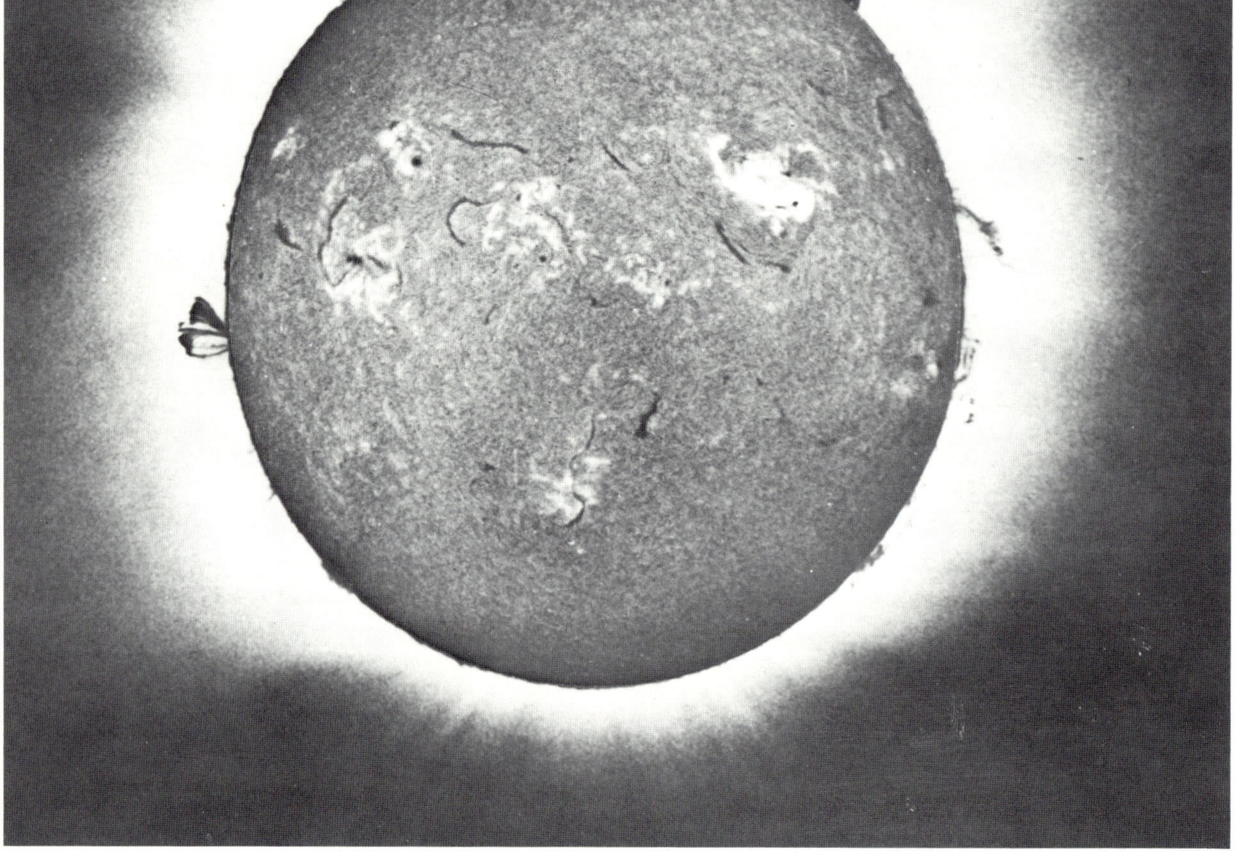

Fig. 28.

THE LIVING SUN

Lockheed Solar Observatory
Lockheed Aircraft Corporation

Fig. 29. Why do the gasses and particles that make up the tail of a comet stay brilliant so long before they fade away? The material on the area of the comet facing the Sun is heated into a gas by radiation pressure from the Sun. The gases are created in the vacuum of space so cool very slowly in the wake of the comet. The cooling allows the gases to condense into particles of dust which are eventually captured by the planets. **Mount Wilson and Palomar Observatories**

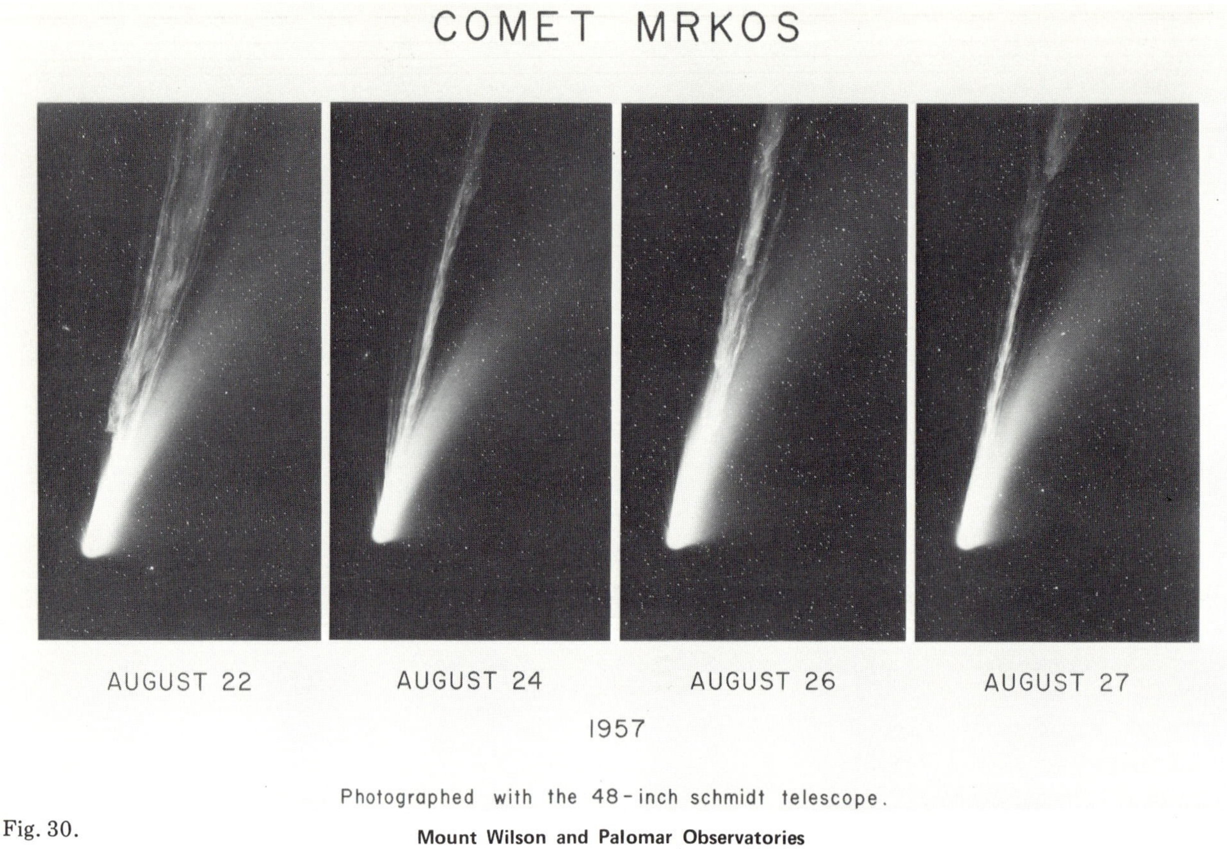
COMET MRKOS
AUGUST 22
AUGUST 24
AUGUST 26
AUGUST 27
1957
Photographed with the 48-inch schmidt telescope.
Mount Wilson and Palomar Observatories

Fig. 30.

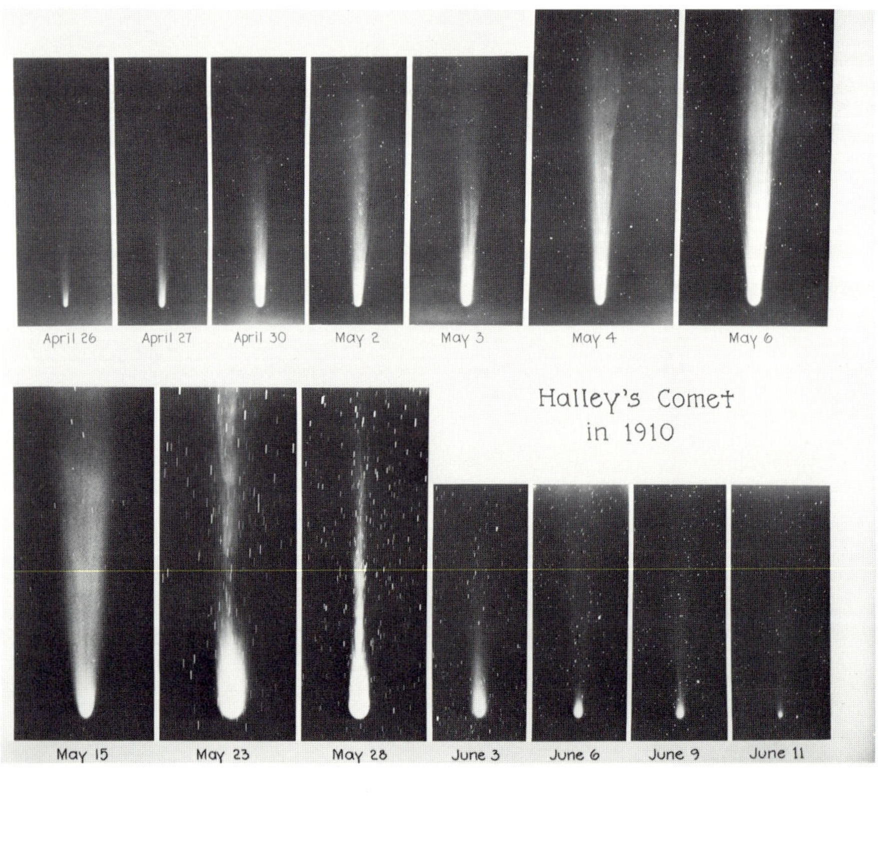
April 26
April 27
April 30
May 2
May 3
May 4
May 6
Halley's Comet
in 1910
May 15
May 23
May 28
June 3
June 6
June 9
June 11

Fig. 31.

Structure of the Planets

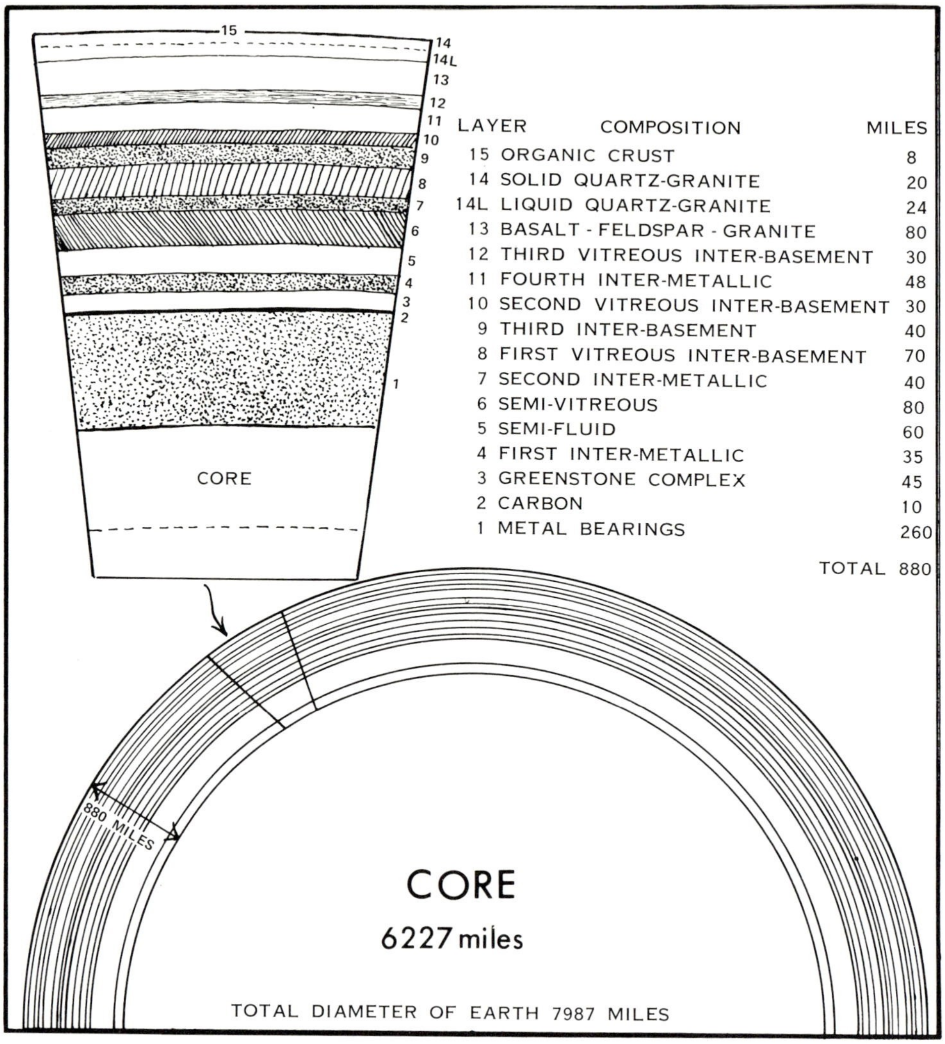

Fig. 32. This figure depicts the various layers of the Earth and their compositions. Present theories indicate a nickel-iron core, but this does not account for the layer, or core, that creates the interior heats of the Earth. Neither does it account for the mechanics of earthquakes.

PRE-CATASTROPHE EARTH

At that time the solid strata of the Earth extended to a depth of about 8 miles. This complex series was mainly composed of the "organic" layers beginning with the Cretaceous down through the Precambrian deposits. Between the 8th and the 52nd mile, there existed a layer of viscous materials composed of quartz and the granites. Below this was a layer of basalt and other similar materials 80 miles thick. From this point to the core the layers were as follows:

* Third Vitreous Inter-basement layer, which was 30 miles thick.
* Fourth Inter-metallic layer, which was 48 miles thick.
* Second Vitreous Inter-basement layer, which is 30 miles thick.
* Third Inter-metallic layer, which was about 40 miles thick.
* First Vitreous Inter-basement layer, which is 70 miles thick.
* Second Inter-metallic layer which was 40 miles thick.
* Semi-vitreous layer, which is 80 miles thick.
* Semi-fluid materials, a layer which was 60 miles thick.
* The first Inter-metallic layer which was 35 miles thick.
* The Carbon layer, which was 10 miles thick.
* The last layer, that which overlies the core of the Earth and is the "bearings" of the crust upon which the outer layers float, was 260 miles thick.

The total of all these layers added up to approximately 880 miles of very mobile crust. (This is most important to remember in order to understand the mechanics of deformation and reformation of the solar bodies.)

Chapter 2

STRUCTURE OF THE PLANETS

Each of the planets is built differently, since each planet's structure is based upon the mode and sequence of its creation. The first planet created was Neptune. It was pulled from the upper or outermost layer of the Sun's atmosphere and elementsphere and is therefore composed mostly of the calcium, silicas and gases. Neptune was quickly followed by Uranus which is also composed of materials from the Sun's outer atmosphere of elements and the silicas, as well as some materials from deeper within the elementsphere. Saturn and Jupiter were both formed at about the same time, of approximately the same amounts of materials. They, too, are mostly compounds of the lighter elements but possess small metallic shells and radioactive cores which are almost absent on Neptune and Uranus.

Pluto, whose original position was between Mars and Jupiter, came from the Sun shortly after Jupiter. Its material contained about equal amounts of rock and metal forming substances as well as an average amount of the atmospheric gases.

Mars followed quickly after Pluto but its material contained higher ratios of metals (mostly in the iron group) and less amounts of the silicates and even less of the Sun's atmosphere. (A large amount of water condensed and collected on the surface after the crust hardened.) Mars was formed just moments after the passing star crossed point "X" in Fig. 4, page 22.

Earth was pulled from the Sun a short time after Mars. But the material came from a deeper region within the elementsphere and so contained a much higher ratio of metal and rock-forming elements and less of the gaseous elements (Fig. 4). This is why the Earth moves faster and closer to the Sun than does Mars.

Venus followed the Earth and though possessing approximately the same volume of material it will be discovered to have a greater density,* which indicates that Venus contains metals of much higher specific gravity such as: nickel, cobalt, tungsten, gold and silver, orginating from near the base of the Sun's elementsphere (just above the main energy emitting surface). Venus also contained less of the Sun's atmospheric materials than did the Earth. It is almost definite that the rock layers on Venus are very thin compared to the metal layers.

Mercury was the last to be pulled away from the Sun and is almost totally metallic. This is indicated by its speed around the Sun and its close proximity.

It must be clearly understood that during the process of the extraction of the planets, all materials were pulled from nearly the same area of the Sun. The extractions were in such rapid order that the groups came from progressively deeper and deeper levels of this one area. Gases and liquids surrounding the "cavity" could not move quickly enough to fill the voids being created in the Sun's

*** Later data will bear this out.**

atmosphere and elementsphere. This is why the planets' compositions grade down from the lighter elements to the heavier, and why atmospheric specific gravities of each planet are different.

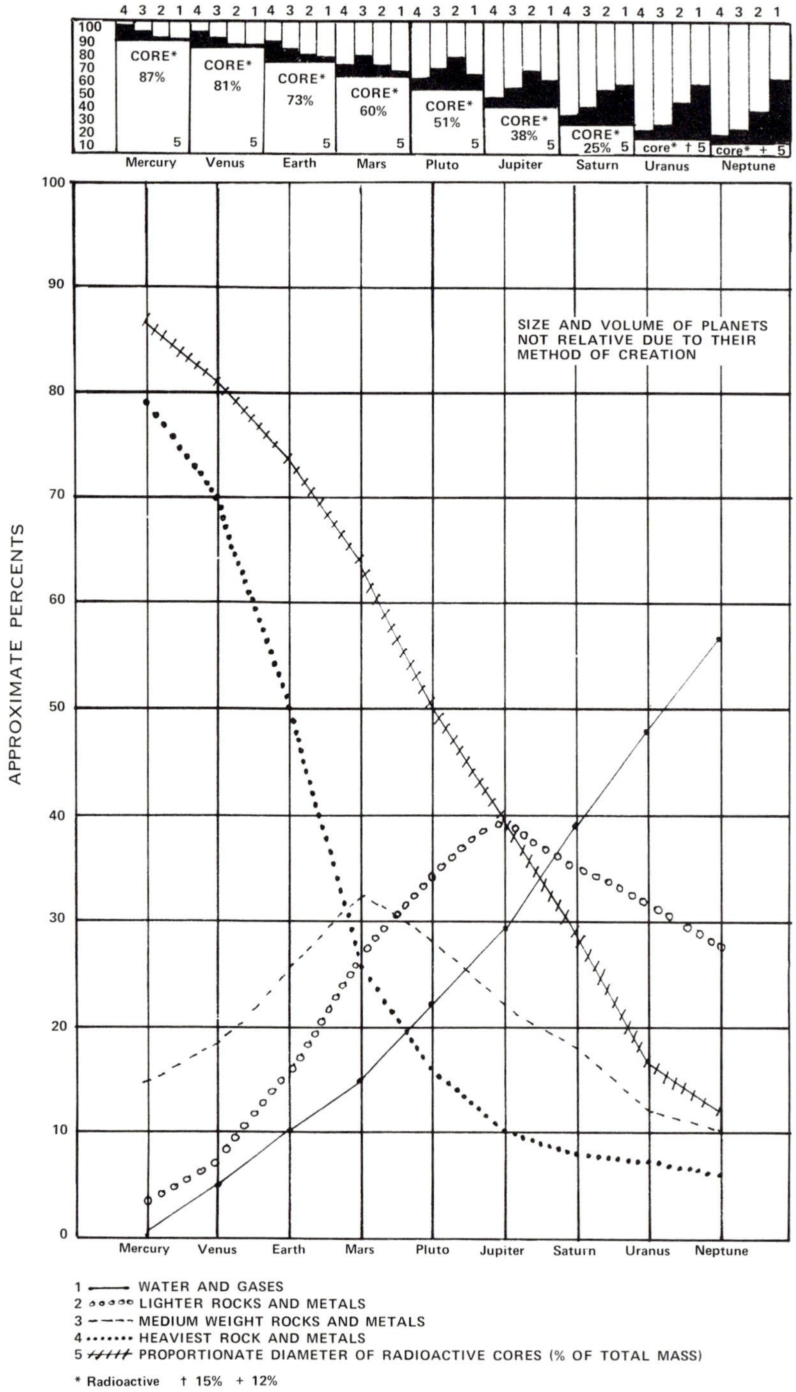

Fig. 33. Ratio and distribution of materials within the planets.

Layer	Element	Symbol	Atomic Number	Specific Gravity
1 (260 MILES)	Osmium	Os	76	22.48
	Iridium	Ir	77	22.421
	Platinum	Pt	78	21.45
	Rhenium	Re	75	20.53
	Wolframium (Tungsten)	W	74	19.3
	Uranium	U	92	18.7
	Tantalum	Ta	73	16.6
	Thorium	Th	90	11.2
	Molybdenum	Mo	42	10.2
	Radon	Rn	86	9.73
	Ruthenium	Ru	44	8.6
	Xenon	Xe	54	5.851
	Radium	Ra	88	5.79
	Erbium	Er	68	4.77
	Krypton	Kr	36	3.708
	Argon	A	18	1.784
	Nitrogen	N	7	1.251
	Neon	Ne	10	.90
	Helium	He	2	.1785
	Hydrogen	H	1	.0899
	Polonium	Po	84	....
	Francium	Fr	87	....
	Actinium	Ac	89	....
	Protactium	Pa	91	....
2 (10 MI.)	Carbon	C	6	2.15
3 (45 MILES)	Hafnium	Hf	72	13.3
	Palladium	Pd	46	11.97
	Neodymium	Nd	60	6.9
	Praseodymium	Pr	59	6.5
	Xenon	Xe	54	5.851
	Iodine	I	53	4.93
	Flourine	F	9	1.69
	Thulium	Tm	69	....
	Ytterbium	Yb	70	....
4 (35 MILES)	Copper	Cu	29	8.92
	Nickel	Ni	28	8.9
	Cobalt	Co	27	8.9
	Tin	Sn	50	7.28
	Chromium	Cr	24	6.92
	Arsenic	As	33	5.727
	Barium	Ba	56	3.5
5 (60 MILES)	Niobium (Columbium)	Cb	41	8.55
	Dysprosium	Dy	66	7.81
	Neodymium	Nd	60	6.9
	Zirconium	Zr	40	6.4
	Lanthanum	La	57	6.15
	Titanium	Ti	22	4.5
	Aluminum	Al	13	2.71
	Hydrogen	H	1	.0899
	Promethium	Pm	61	....
	Holmium	Ho	67	....
6 (80 MILES)	Thallium	Ti	81	11.85
	Gallium	Ga	31	5.903
	Germanium	Ge	32	5.39
	Iodine	I	53	4.93
	Strontium	Sr	38	2.60
	Boron	B	5	2.3
	Technetium (Masurium)	Tc	43	....
	Lutecium	Lu	71	....
	Astatine	At	85	....
	Francium	Fr	87	....
7 (40 MILES)	Gold	Au	79	19.3
	Silver	Ag	47	10.5
	Iron	Fe	26	7.86
	Manganese	Mn	25	7.2
	Vanadium	V	23	5.866
	Silicon	Si	14	2.14
8 (70 MILES)	Samarium	Sa	62	7.7
	Europium	Eu	63	7.42
	Cerium	Ce	58	6.9
	Antimony	Sb	51	6.684
	Tellurium	Te	52	6.25
	Radium	Ra	88	5.79
	Barium	Ba	56	3.5
	Strontium	Sr	38	2.60
	Cesium	Cs	55	1.9
	Magnesium	Mg	12	1.74
	Calcium	Ca	20	1.55
	Oxygen	O	8	1.429
	Indium	In	49	....
	Polonium	Po	84	....
	Francium	Fr	87	....
9 (40 MILES)	Bismuth	Bi	83	9.8
	Cadmium	Cd	48	8.642
	Tin	Sn	50	7.28
	Beryllium	Be	4	1.85
	Sodium	Na	11	.97
	Lithium	Li	3	.534
10 (30 MILES)	Rhodium	Rh	45	12.1
	Yttrium	Y	39	5.51
	Iodine	I	53	4.93
	Gadolinium	Ga	64	4.52
	Terbium	Tb	65	4.35
	Scandium	Sc	21	2.51
	Rubidium	Rb	37	1.532
	Oxygen	O	8	1.429
11 (48 MILES)	Mercury	Mg	80	13.546
	Lead	Pb	82	11.344
	Zinc	Zn	30	7.14
	Selenium	Se	34	4.26
	Sulphur	S	16	2.07
	Flourine	F	9	1.69
12 (30 MILES)	Rhodium	Rh	45	12.1
	Yttrium	Y	39	5.51
	Iodine	I	53	4.93
	Bromine	Br	35	2.928
	Scandium	Sc	21	2.51
	Phosphorus	P	15	1.82
	Magnesium	Mg	12	1.74
	Rubidium	Rb	37	1.532
	Oxygen	O	8	1.429
	Potassium	K	19	.86
13 (80 MILES)	Bromine	Br	35	2.928
	Phosphorous	P	15	1.82
	Magnesium	Mg	12	1.74
	Oxygen	O	8	1.429
	Potassium	K	10	.86
14 (44 MILES)	Chlorine	Cl	17	3.214
	Silicon	Si	14	2.14
	Flourine	F	9	1.69
	Calcium	Ca	20	1.55
	Oxygen	O	8	1.429
15 (SEDIMENTARY CRUST, 8 MILES)	Chlorine	Cl	17	3.214
	Aluminum	Al	13	2.71
	Boron	B	5	2.3
	Silicon	Si	14	2.14
	Sulphur	S	16	2.07
	Phosphorous	P	15	1.82
	Carbon	C	6	1.8
	Magnesium	Mg	12	1.74
	Flourine	F	9	1.69
	Oxygen	O	8	1.429
	Nitrogen	N	7	1.251
	Calcium	Ca	20	1.55
	Sodium	Na	11	.97
	Neon	Ne	10	.90
	Potassium	K	19	.86
ATMOSPHERE (660 MILES)	Krypton	Kr	36	3.708
	Argon	A	18	1.784
	Oxygen	O	8	1.429
	Nitrogen	N	7	1.251
	Helium	He	2	.1785
	Neon	Ne	10	.90
	Hydrogen	H	1	.0899

TABLE OF THE ELEMENTS

Fig. 34. Table shows the elements as they lay in the Earth in the order in which they finally settled after their order of specific gravity. Specific gravity — not boiling points — governed the final position taken by the elements. Certain elements would seem not to belong in these layers. This is due to their specific gravities and cooling —being "trapped" in "out of place" positions. Others, as in layer 1 are new, being continually created and adding to this layer.

Fig. 35.

THE PRE-CATASTROPHE EARTH

Catastrophe and Chaos

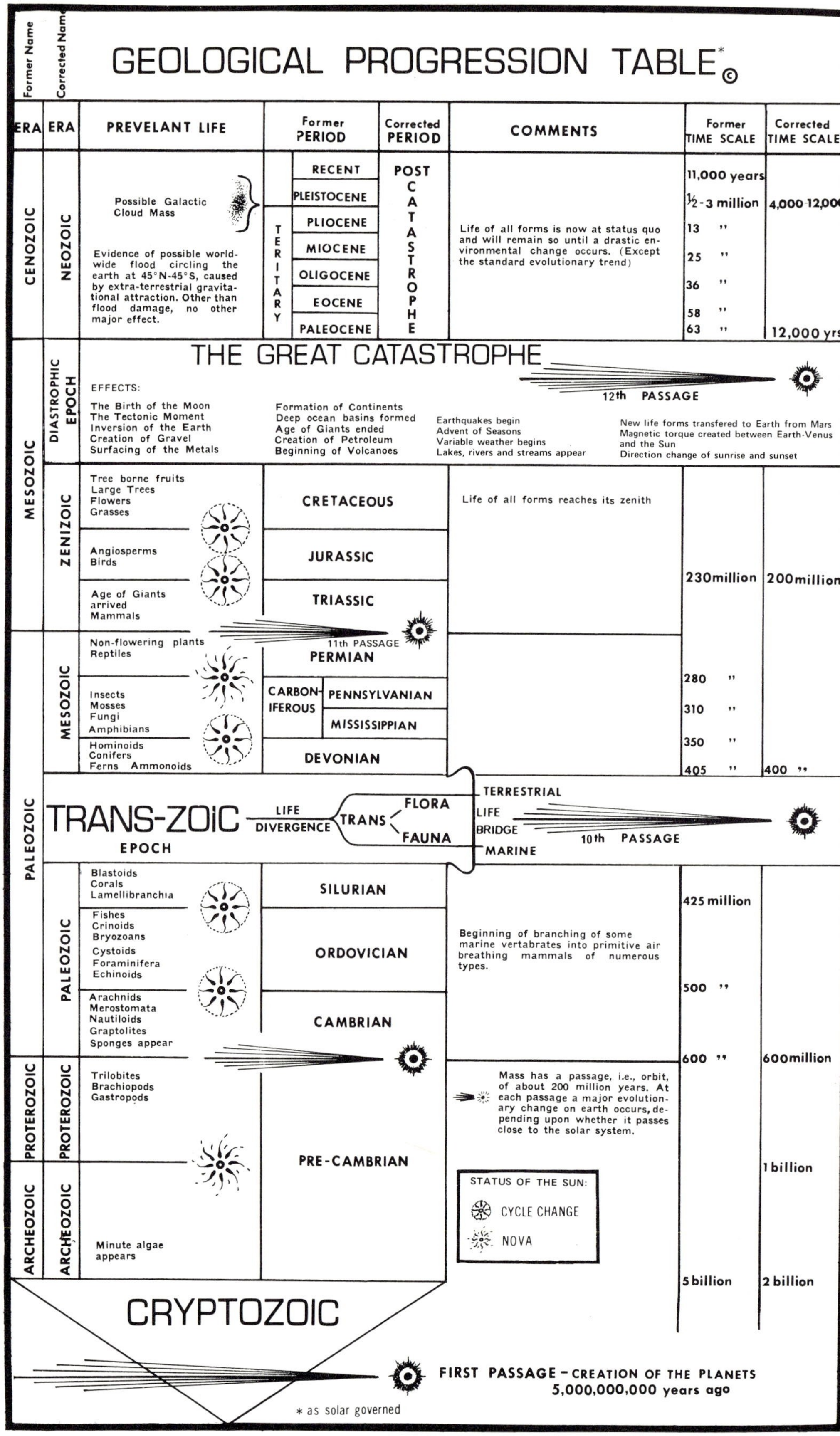

Former Name
Corrected Name
GEOLOGICAL PROGRESSION TABLE* ©
ERA
ERA
PREVELANT LIFE
Former PERIOD
Corrected PERIOD
COMMENTS
Former TIME SCALE
Corrected TIME SCALE
CENOZOIC
NEOZOIC
Possible Galactic Cloud Mass
Evidence of possible world-wide flood circling the earth at 45°N-45°S, caused by extra-terrestrial gravitational attraction. Other than flood damage, no other major effect.
TERTIARY
RECENT
PLEISTOCENE
PLIOCENE
MIOCENE
OLIGOCENE
EOCENE
PALEOCENE
POST CATASTROPHE
Life of all forms is now at status quo and will remain so until a drastic environmental change occurs. (Except the standard evolutionary trend)
11,000 years
½-3 million
13 "
25 "
36 "
58 "
63 "
4,000-12,000
12,000 yrs
MESOZOIC
DIASTROPHIC EPOCH
THE GREAT CATASTROPHE
12th PASSAGE
EFFECTS:
The Birth of the Moon
The Tectonic Moment
Inversion of the Earth
Creation of Gravel
Surfacing of the Metals
Formation of Continents
Deep ocean basins formed
Age of Giants ended
Creation of Petroleum
Beginning of Volcanoes
Earthquakes begin
Advent of Seasons
Variable weather begins
Lakes, rivers and streams appear
New life forms transfered to Earth from Mars
Magnetic torque created between Earth-Venus and the Sun
Direction change of sunrise and sunset
ZENIZOIC
Tree borne fruits
Large Trees
Flowers
Grasses
Angiosperms
Birds
Age of Giants arrived
Mammals
CRETACEOUS
JURASSIC
TRIASSIC
Life of all forms reaches its zenith
230million
200million
PALEOZOIC
MESOZOIC
Non-flowering plants
Reptiles
11th PASSAGE
PERMIAN
Insects
Mosses
Fungi
Amphibians
Hominoids
Conifers
Ferns Ammonoids
CARBONIFEROUS
PENNSYLVANIAN
MISSISSIPPIAN
DEVONIAN
280 "
310 "
350 "
405 "
400 "
TRANS-ZOIC EPOCH
LIFE DIVERGENCE
TRANS
FLORA
FAUNA
TERRESTRIAL
LIFE
BRIDGE
MARINE
10th PASSAGE
PALEOZOIC
Blastoids
Corals
Lamellibranchia
Fishes
Crinoids
Bryozoans
Cystoids
Foraminifera
Echinoids
Arachnids
Merostomata
Nautiloids
Graptolites
Sponges appear
SILURIAN
ORDOVICIAN
CAMBRIAN
Beginning of branching of some marine vartebrates into primitive air breathing mammals of numerous types.
425 million
500 "
600 "
600million
PROTEROZOIC
PROTEROZOIC
Trilobites
Brachiopods
Gastropods
PRE-CAMBRIAN
Mass has a passage, i.e., orbit, of about 200 million years. At each passage a major evolutionary change on earth occurs, depending upon whether it passes close to the solar system.
1 billion
ARCHEOZOIC
ARCHEOZOIC
Minute algae appears
STATUS OF THE SUN:
CYCLE CHANGE
NOVA
5 billion
2 billion
CRYPTOZOIC
FIRST PASSAGE - CREATION OF THE PLANETS
5,000,000,000 years ago
* as solar governed

Fig. 36.

Chapter 3

CATASTROPHE AND CHAOS

Each time, in their 200 million year orbits around the galaxy, our Sun and the other star, which we will call, Mazzaroth, meet; the encounter causes a major change within the solar system. On Earth we find the evidence of these changes in the different geologic Eras and/or Epochs.* This evidence is revealed in the rock formations, their compositions, thickness of deposits and areas covered. Fossil records in and between these strata tell the 5 to 10 billion year story of the Sun's activity. The strata-covered fossils graphically tell when one cycle on the Sun ended and another began. If the cycle change was sudden and of great magnitude, the evidence of a nova on the Sun is told in the rocks by their thickness and the type of material deposited as well as the lack of fossils.

The periodic passage of the star close to the Sun has caused major effects on the Sun, which either changed the depth of the elementsphere or caused, through gravitational effect, cyclic energy emission changes which produced new forms of radiation environment on the earth.

Whenever a nova has occurred on the Sun as it did during the Precambrian period, and the period which ended the Carboniferous and began the Permian, great changes occurred on the Earth. New forms of life, an entirely new environment brought about by drastic solar radiation changes, resulted in entirely different branching of various forms of life.

Due to a cycle change on the Sun the Ordovician period began with the fishes evolving late in the cycle. The corals were the result of a cycle change which began the Silurian, and in the mid-Devonian times the conifers and the ferns – and man – appeared.

All these new forms of life evolved only because the Sun had changed its *type* of energy emissions, resulting in a different form of environment within the Earth's atmosphere, water, and effect on plant growth. The blossom of every flower on Earth today is the result of a special adaptation and response to a specific range of solar energy.

The passage of the star 200 million years ago brought about the evolvement of mammals and the arrival of the giants. In a short period of time this cycle change brought forth the birds and the angiosperms. This period of the Jurassic lasted only a few million years and another cycle change brought in a new period called the Cretaceous, during which evolved the grasses, large trees, flowers, and the tree borne fruits.

The last meeting between our Sun and the star was approximately 12,000 years ago† and brought to an abrupt end the Cretaceous period. The Earth was a much different planet 12,000 years ago from what it is today.

* See Geological Progression Table

† See Appendix page 193

The Earth was covered by shallow seas and low land areas. It was a nearly smooth, swampy, humid globe covered from pole to pole with fish, birds, reptiles, mammals – large and small (including man) – and great masses of plant life. That the Earth was thickly populated by animals and marine life is evidenced by the great quantities of their remains in petroleum, coal beds, and fossil beds. The amount of plant life required to support the countless numbers and kinds of animals would of necessity fill the shallow seas and cover the lands with a dense growth. Since there were no seasons at that time (Fig. 37), the Earth, blanketed by a dense atmosphere, was like a humid greenhouse. Growth was constant and uninterrupted. Life of every form was at its pinnacle.

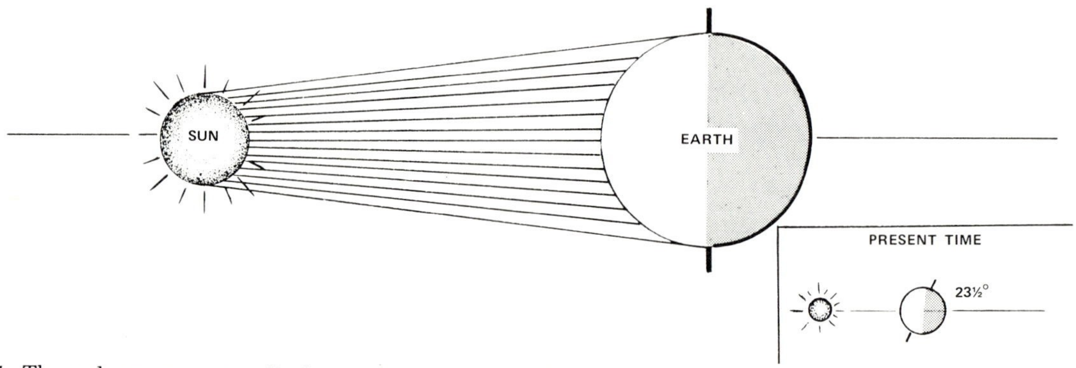

Fig. 37. The poles were perpendicular to the plane of the Earth's orbit, allowing unvarying radiation and illumination from the Sun at all times of the year.

12,000 years ago – oceans, rivers, mountains, deep valleys, rock, flint, metals and petroleum were non-existent upon the face of the Earth. Since there were no deep oceans, there were no beaches no sand or gravel. There was no rain, no snow, no lightning or thunder, and no wind. Primitive man had never seen the Sun or the stars because of the deep dense atmospheric canopy – and the Moon did not exist.

Then the solar system, moving in its great orbit along the edge of the galaxy, was again encountered by the other star. At this last meeting many radical changes were brought about among the planets. The intruding star (much larger than our Sun) came closer than it had the previous ten times. This time it passed right through the solar system. It was traveling approximately 1, 054, 000 miles per hour faster that the Sun was moving. In about 60 weeks the star passed completely through the Sun's planetary system leaving catastrophe and chaos in its wake (Fig. 41 and 42).

The star entered the solar system at 17½° angle to the plane of the planets' orbits. It passed very close to Mars, Jupiter, Saturn, Uranus and Neptune; but it nearly struck Pluto, which was then occupying an orbital position between Mars and Jupiter (Fig. 42, pages 54 and 55).

Because of its great size, had the star collided with any of the planets it would have shattered them. There would have been no residue from such a collision since the gravity of the star would have pulled all materials to its surface immediately. Though it did not come in direct contact with any member of the solar system, it did do considerable damage. The great gravitational attraction of the star pulled upon each planet and the Sun and dragged great trails of gas away from them. These gases streamed out into space followed by thousands of crustal sections and huge writhing globules of molten metal from Earth, Mars, and some from Jupiter.

As the attraction of the star waned, the gravity of the Sun pulled all this material streaming after the star into the orbit vacated by the nearly destroyed Pluto. This material now fills what is known as the Asteroid belt (Fig. 38). The Asteroids are debris from the various planets but mainly of the crust of Pluto. There are literally millions of small fragments flying through space in this particular orbit. The Asteroids move around the Sun in the same direction as the planets and in the same direction and angle of the passing star that created them.

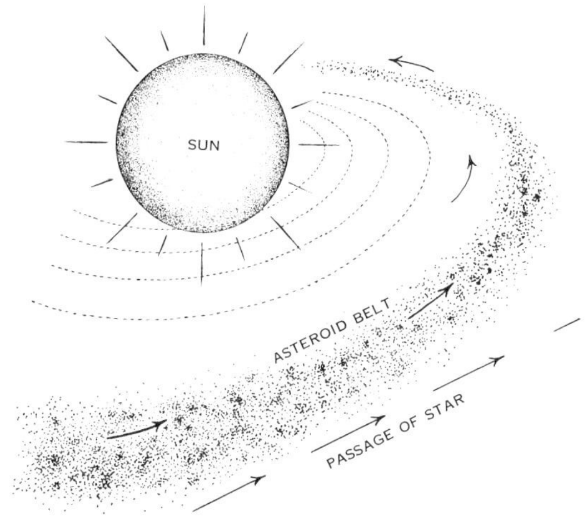

Fig. 38.

The path of the star is easily traced by the tilt of Saturn's rings, the orbital angle of each planet's satellites, and most important of all – the direction in which the different satellites move in their orbits around their planets.

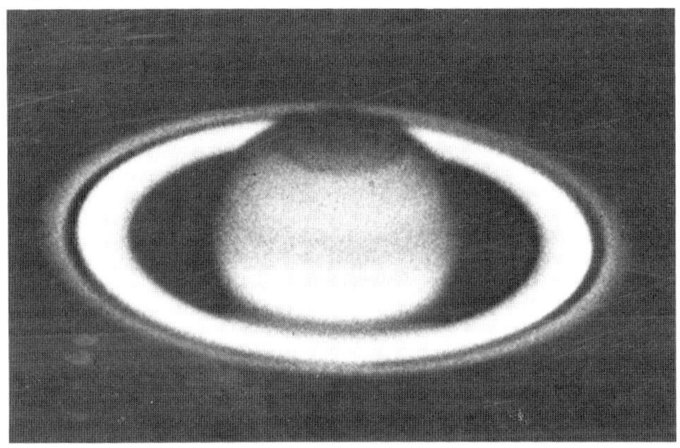
Fig. 39 Mount Wilson and Palomar Observatories

For instance, the satellites belonging to the Earth and Mars move in the same direction as their planet's rotations, while the majority of the satellites on the *outer* side of the Asteroid belt move in a retrograde direction, one opposite to their planet's rotation. This is additional evidence that the star passed through the solar system on this path.

Some satellites move retrograde to others revolving around the same planet (Jupiter, Uranus and Saturn). These satellites were pulled from the planet and set into motion as the star approached the planet, then as the star passed by, the other masses were pulled away from the planet to circle it in the opposite direction from the first group. Every planet was affected in some manner by the stellar encounter, some more than others. The amount of damage each planet suffered depended upon where it was located in the solar system at the time and the individual construction unique to the planet.

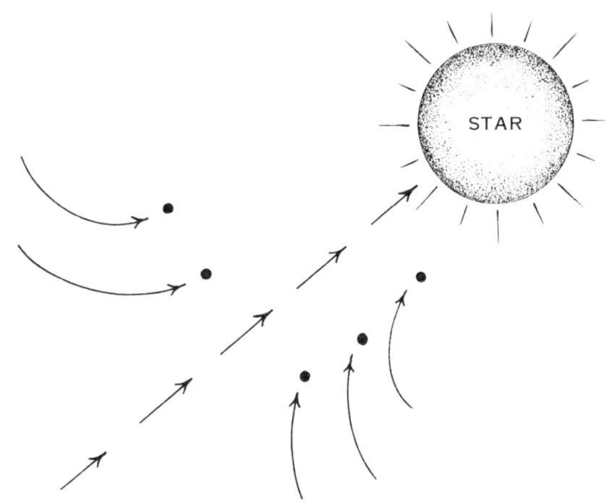

Fig. 40.

During the star's passage, Mercury was on the opposite side of the Sun and probably suffered very little damage except for minor disruption of its thin crust with some intrusion into the surface layers by the thin liquid granite layer below.

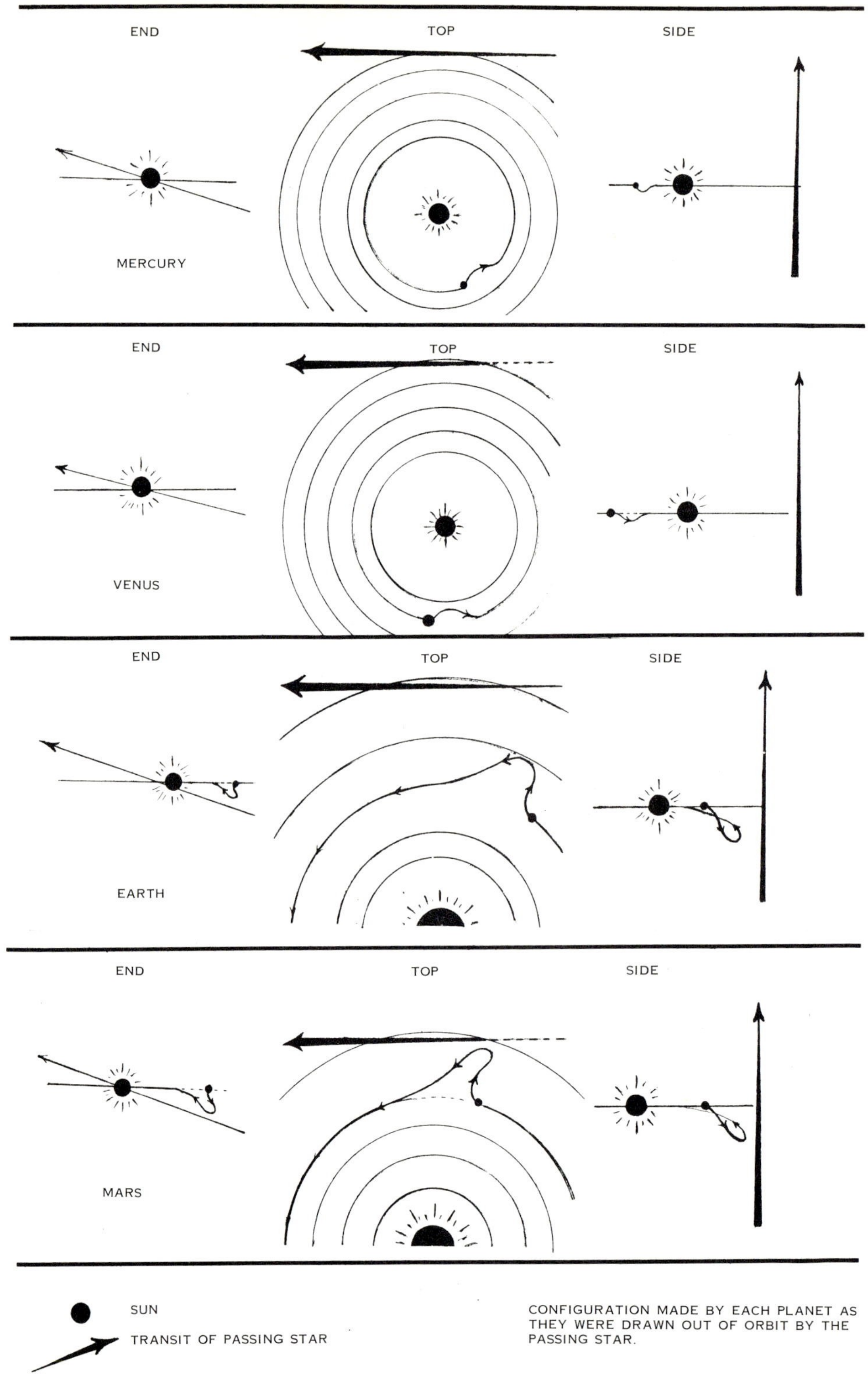

Fig. 41 and 42. These figures show the path of the star through the solar system — both top and side views. The figures show the positions of the planets at the time and the angle of the star's transit through the plane of the orbits.

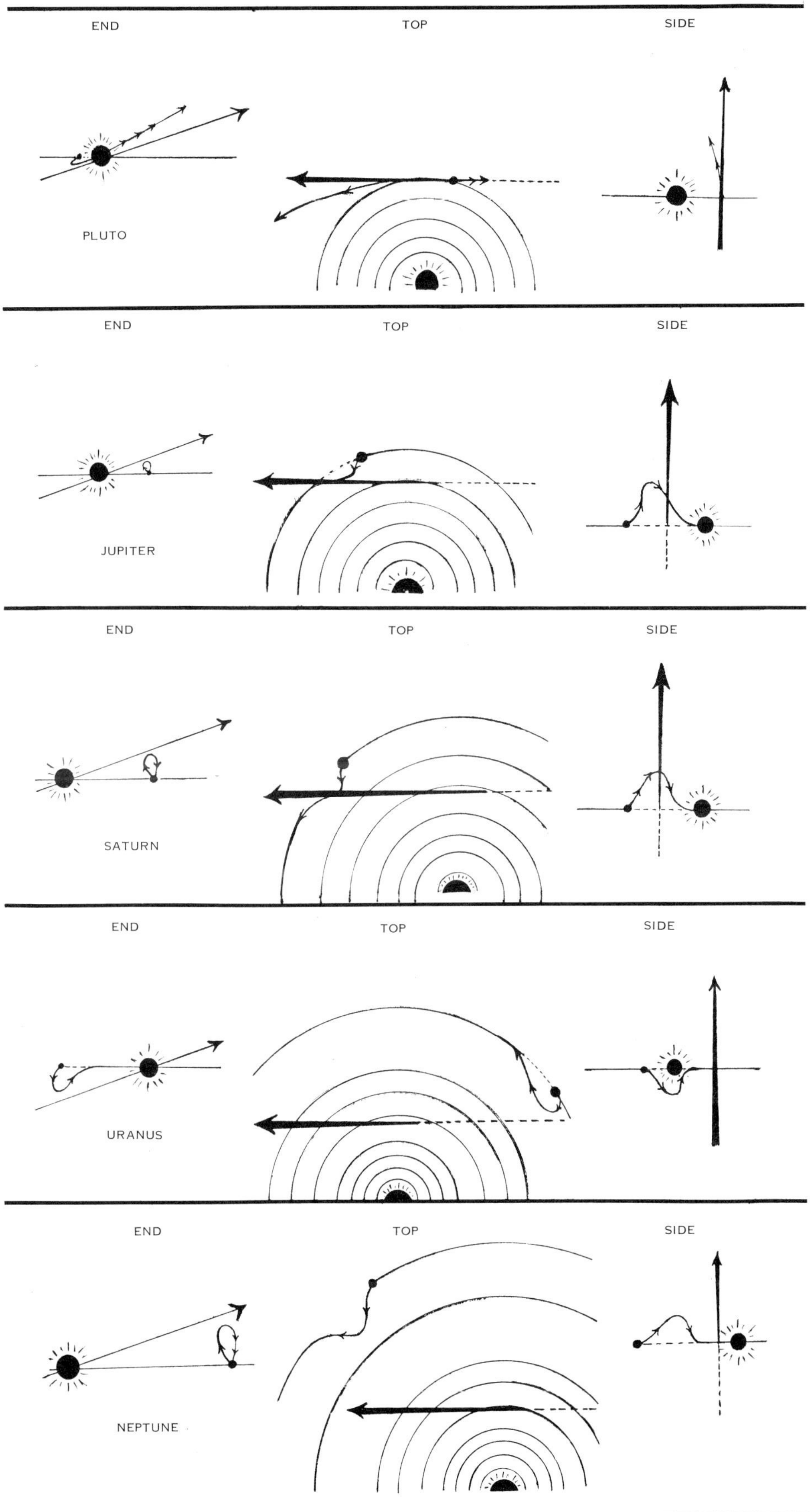
END
TOP
SIDE
PLUTO
END
TOP
SIDE
JUPITER
END
TOP
SIDE
SATURN
END
TOP
SIDE
URANUS
END
TOP
SIDE
NEPTUNE

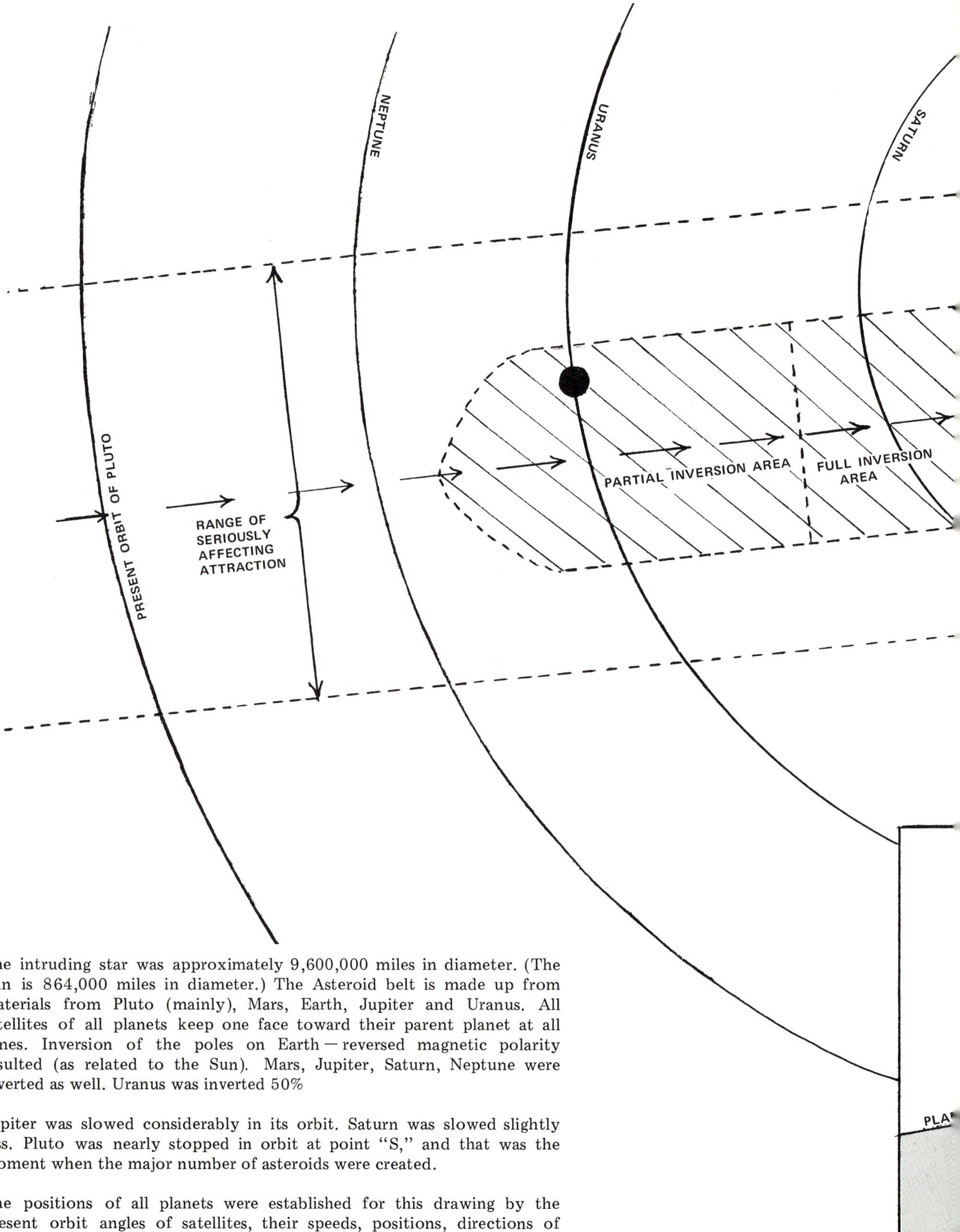

The intruding star was approximately 9,600,000 miles in diameter. (The Sun is 864,000 miles in diameter.) The Asteroid belt is made up from materials from Pluto (mainly), Mars, Earth, Jupiter and Uranus. All satellites of all planets keep one face toward their parent planet at all times. Inversion of the poles on Earth — reversed magnetic polarity resulted (as related to the Sun). Mars, Jupiter, Saturn, Neptune were inverted as well. Uranus was inverted 50%

Jupiter was slowed considerably in its orbit. Saturn was slowed slightly less. Pluto was nearly stopped in orbit at point "S," and that was the moment when the major number of asteroids were created.

The positions of all planets were established for this drawing by the present orbit angles of satellites, their speeds, positions, directions of travel, tilt or planetary axis, magnetic field status and gravitational values.

Fig. 42.

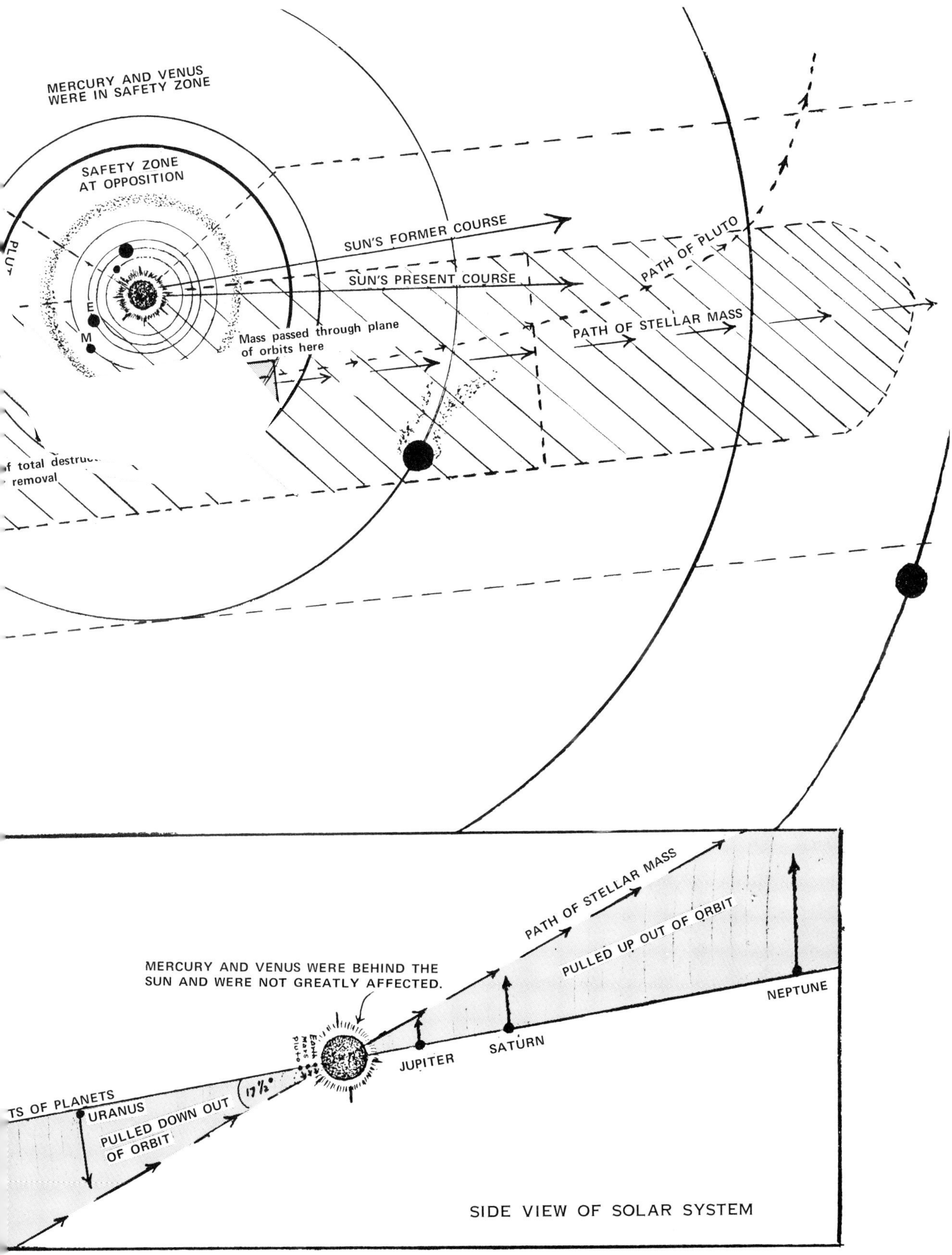
MERCURY AND VENUS WERE IN SAFETY ZONE
SAFETY ZONE AT OPPOSITION
E
M
SUN'S FORMER COURSE
SUN'S PRESENT COURSE
PATH OF PLUTO
PATH OF STELLAR MASS
Mass passed through plane of orbits here
removal
PATH OF STELLAR MASS
PULLED UP OUT OF ORBIT
MERCURY AND VENUS WERE BEHIND THE SUN AND WERE NOT GREATLY AFFECTED.
NEPTUNE
SATURN
JUPITER
Earth
Mars
Pluto
17½°
URANUS
PULLED DOWN OUT OF ORBIT
SIDE VIEW OF SOLAR SYSTEM

Venus was also on the opposite side of the Sun during opposition of the star (Fig. 43) and was saved from material damage. Though the planet lost no material, its outer crust was surely wrecked. The surging liquid metal layers would have caused the surface to buckle and crack. The quartz-granite layer is but one-third as thick as the same layer on Earth.

Venus lost most of its water and atmosphere since at the moment of the star's opposition to Earth, Venus was passing from behind the Sun and was exposed to the greater attraction of the Star. Thus, its air and water departed.

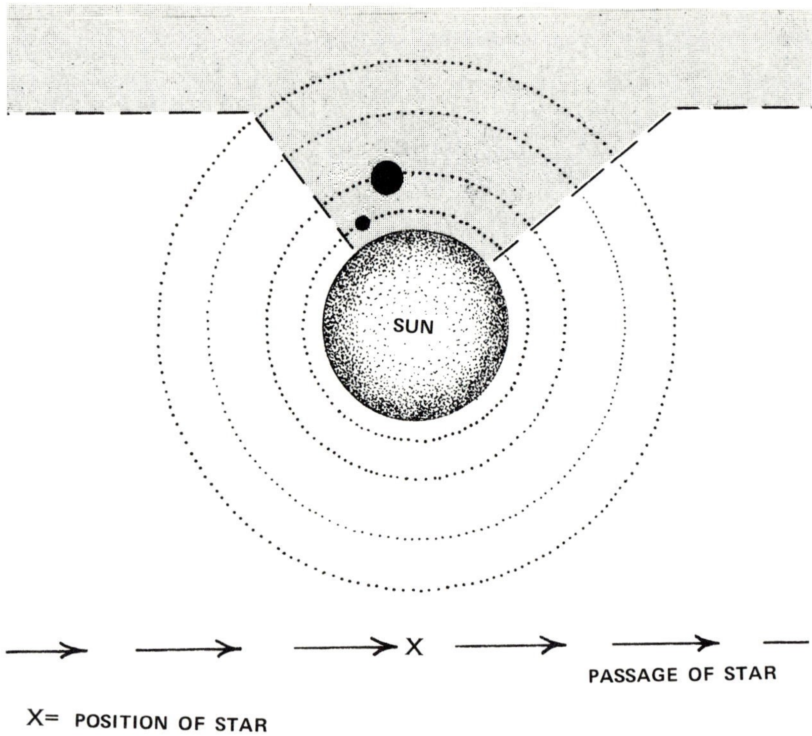

Fig. 43.

However, when the star passed in opposition to the Sun, its attraction pulled large volumes of gases from the Sun which were retrieved by Venus just a few weeks later. Venus now has far more atmosphere than formerly. Since it is similar in size and structure to the Earth,* had Venus been on the same side of the Sun as the Earth, it, too, would have produced a satellite.

Because of Mars' orbital position at the time, it was quite close to the path of the passing star. Mars was forced to relinquish 70% of its crust and practically all of its atmospheric cover, including 90% of its moisture (Fig. 51). Recent discoveries by Mariner and Viking space vehicles to Mars have shown that there have been catastrophic floods and many indications of large volumes of water formerly on Mars.

The axis of Mars is inclined to the plane of its orbit 24 degrees – the same as Earth's – and it now has a diameter about half that of Earth with a surface gravity 4/10ths as strong as expected for gravity. Mars also inverted with over half of its crust stripped away but having a much thicker outer crust and being layered differently from the Earth in the beginning, Mars lost more mass. (See Fig. 183 which shows remaining continental crust.) Much of its mass became asteroids and comets. Although the water is missing, note the continental outlines on the maps. The northern hemisphere of Mars lies several kilometers above the average level of the southern hemisphere which is definite indication that a major portion of Mars' crust and part of its interior fluids were removed. What we now see exposed is actually the continental crust standing above empty ocean basins – the water having been pulled away by the passing star (See Fig. 183, pages 158 and 159).

*** All planets are similar in structure. Only the ratio of layers and the cores are different. The number of layers overlying the cores of each planet varies as does the thickness and content, as well as their ratio of elements.**

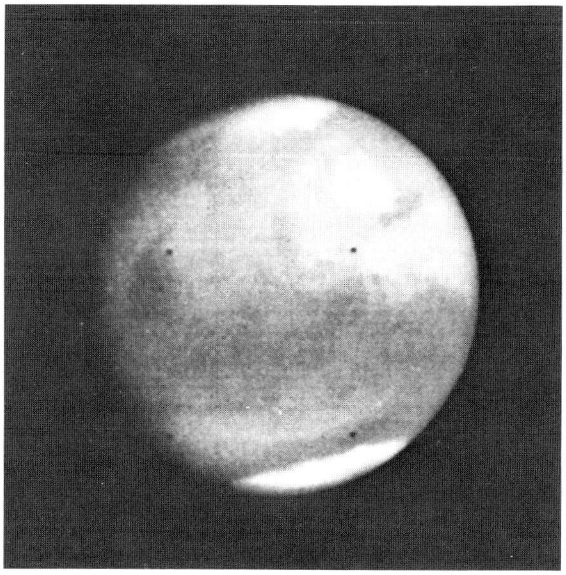

Fig. 44. The planet Mars. NASA

Mars now travels in an elliptical orbit and has recaptured two lost sections which have become its two moons. The total surface of Mars is heavily covered with great quantities of all metals. During the removal of the crust, oceans of metal from all the metallic layers rose to the surface and mixed thoroughly with the various liquid rock layers remaining. As a result the surface of Mars has a strong reddish cast caused by the heavy extrusion of metals which drifted as gases over the surface for many hours after the catastrophe.

Over the past 12,000 years Mars, like the Moon, lacking the necessary ingredients in its atmosphere, has experienced almost no weathering or eroding of its surface. All features instilled at the time of the catastrophe have remained essentially the same.

The great amount of evidence now found for former running water in many places on Mars confirms the sudden removal of its original deep water cover. It is also highly probable that in addition to some marine life, several kinds of plants from Mars were given a new start on Earth. It is quite likely that some of the surface debris pulled away from Mars was drawn to the Earth. This would answer some of the enigmas regarding the origins of some of the forms of life now found on Earth. Certain forms of highly evolved life seem to be without a "past" for evidence of their evolvement is lacking in the fossil record.

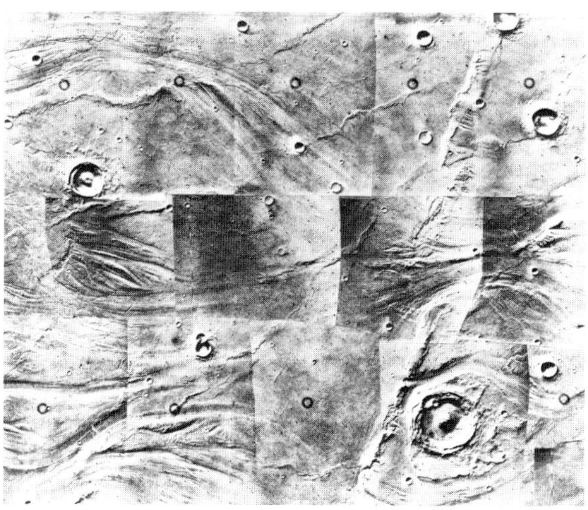

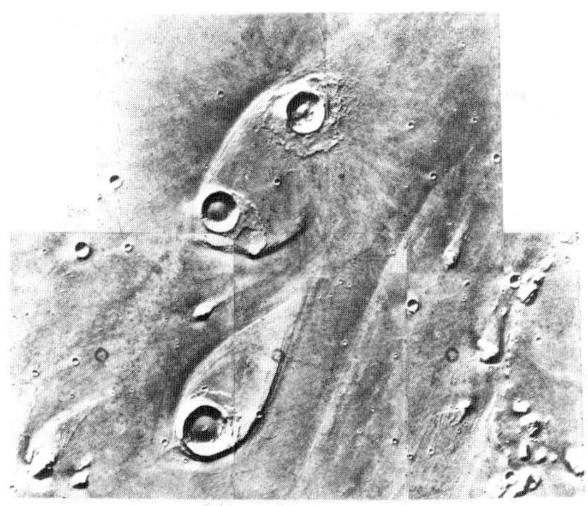

Fig. 45, 46. Flow lines show exodus of water from Mars. All flows are uniformly shallow. NASA

Regarding new forms of life on Earth – the recent arrival of flora and fauna which have no prior evolvement evidence on Earth is most important. Some of these new forms are: photosynthesizing algae and calcareous foraminifera. Other new life forms that arrived from Mars (some possibly from Venus) came as eggs, seeds, and spores. "New fossils" – life forms that have generated since the catastrophe lie within 2 to 6 inches of the Earth's present surface.

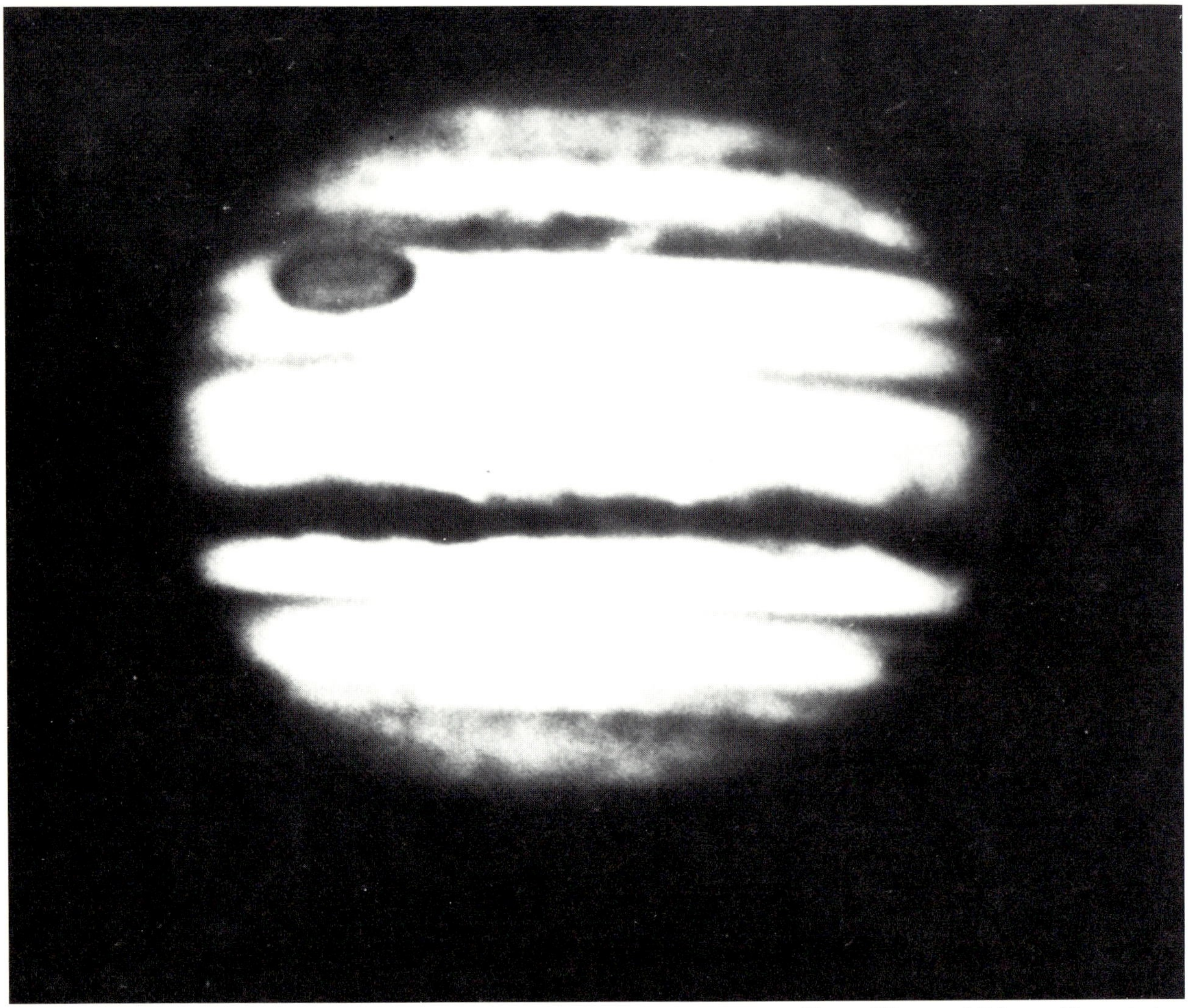

Mount Wilson and Palomar Observatories

Fig. 47. The position of the bulge points to the angle of transit of the star in the vicinity of Jupiter.

THE PLANET JUPITER

Jupiter was caught in a very vulnerable position when the star passed by, and it gave up 14 major sections of its crust. As the star approached, 8 of these sections were pulled from the surface of the planet and as the star passed by, 6 more sections were pulled away in the opposite direction (Fig. 48). All of the sections were recaptured into orbit around Jupiter – 8 moving with the rotation of the planet and 6 moving retrograde.

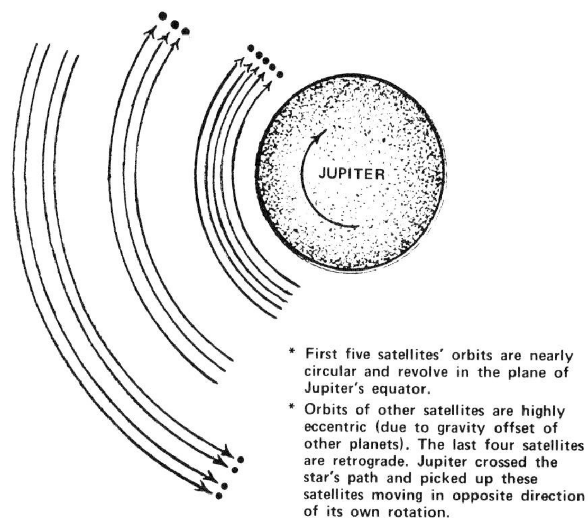

Fig. 48.

I suspect that the great red bulge seen on the surface of Jupiter is a mass of cooled fluid rock that was drawn out of Jupiter by the passing star. It was the last of the materials that attempted to follow the fourteen other masses extracted from Jupiter that became her moons, but the rapidly diminishing gravitational attraction left this great red bulge still attached to its parent. Some have called this an atmospheric disturbance, but no atmospheric anomaly could be so permanent.

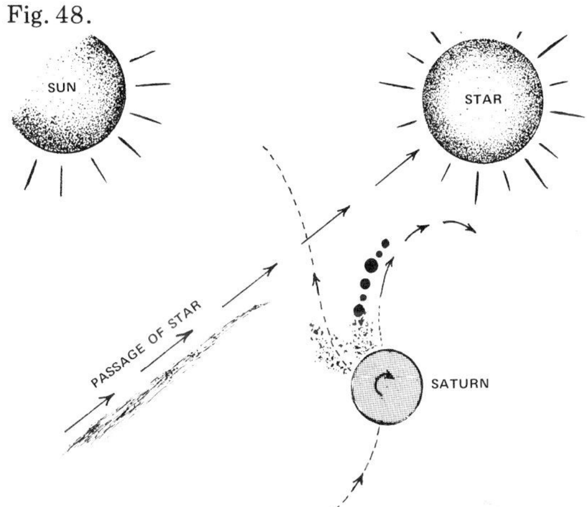

Fig. 49.

Saturn was to the right (with the Sun to the left) and below the path of the star. Saturn lost one section of its crust as the star approached and 8 others with its passing, including a great quantity of material of small dimensions that now moves around the planet in a flat orbital plane. Therefore, Saturn has one satellite moving retrograde to its own rotation and rings of gas and small particles flying at high speeds around its inner influence in diverse directions of orbital motion. Fig. 50 shows how the rings were acquired. A great quantity of the material forming the two moving rings came from material orginating from the other planets.*†

Sixty weeks were required for the star to make its transit through the solar system – a distance of 5,600,000,000 miles (the diameter of Neptune's orbit), thus allowing such planets as Jupiter and Saturn sufficient time to cross the wake of the star as the planets traveled in their orbits. These

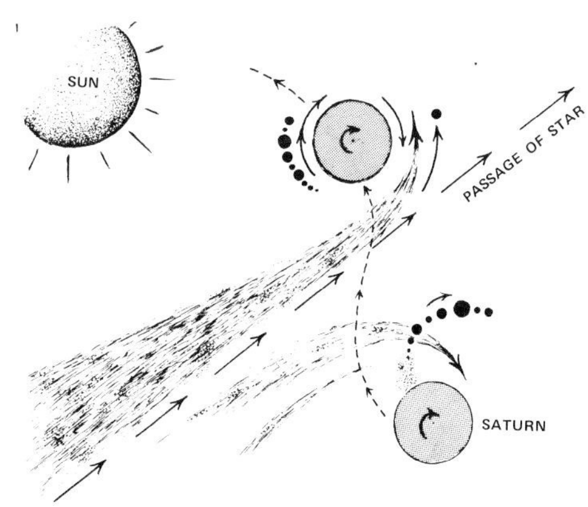

Fig. 50.

* A tenth satellite has been recently discovered by Audouin Dollfus of the Meudon Observatory. It is associated with the rings and was undoubtly captured at the same time.

† This transfer of materials from one planet to another gives evidence to the transfer of large volumes of water from Mars to Earth.

NOTE:

THE 10TH PASSAGE OF THE STAR (MAZZAROTH ENDED THE SILURIAN AGE AND REMOVED ABOUT 3000 FEET OF WATER COVER FROM EARTH

GRAPH SHOWS DEPTH OF WATER COVER ON EACH PLANET BEFORE THE CATASTROPHE – IN THOUSANDS OF FEET.

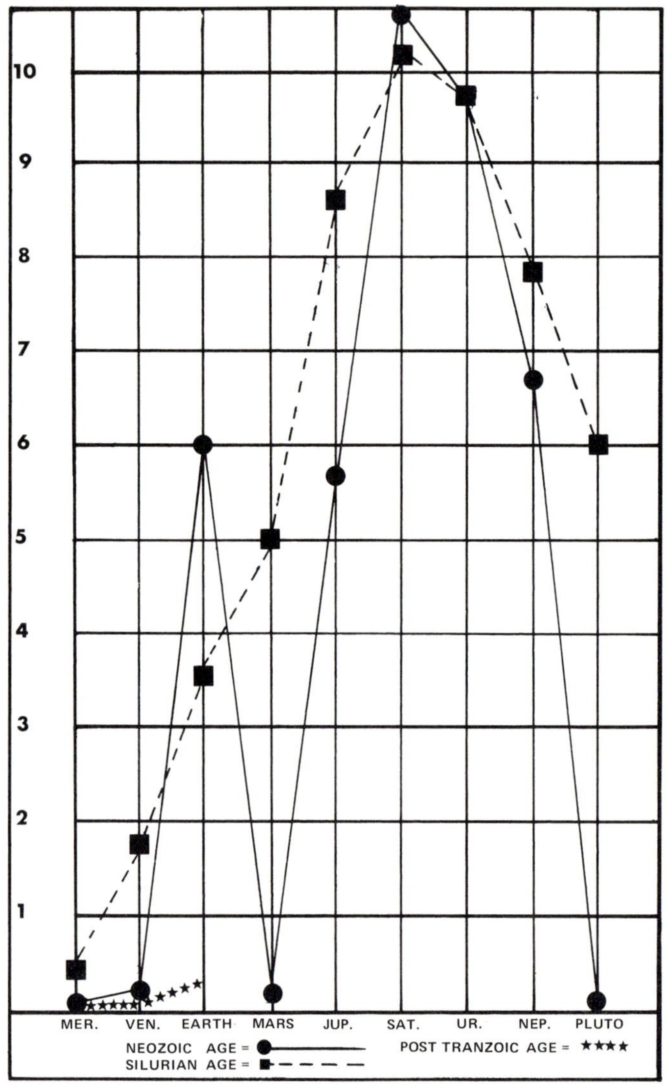

Fig. 51.

If you will refer to the chart on pages 54 and 55 which shows the path of the star, you will note the close proximity of Mars to the path of the star.

Upon approach, the star pulled Mars down and out of its orbit and just before opposition, the gravitational attraction of the star stripped away almost all of the atmosphere and all of the surface water. At opposition, the greater part of the crust was stripped away creating the moons and material for the asteroid belt, comets and meteors.

As the star passed on, it began to duplicate upon earth the affect it had just created on Mars. The atmosphere and oceans of Mars were trailing behind the star but were rapidly falling behind since the speed of the star was too great for this material to accelerate fast enough to catch up. Earth, too, was being pulled down out of orbit, and due to this, it was pulled into the path of the streaking gases and waters of Mars. This almost unbelievable activity is covered in the chapter "The Great Ocean Transfer" in this and the larger work, OUT OF THE GREY MIST" (in preparation).

Therefore, the Earth acquired the atmosphere and waters of Mars simply, yet in an almost unbelievable fashion. Oddly enough the water was "just enough" to cover the continental shelves on the Earth.

The present ratio of water on Earth to the other elements is excessive. The graph shows the original, relatively proper ratio of water at the time of the planets' creation and shows that the depth of water on Earth should have been 6,000 feet during the Silurian era and 200 to 300 feet deep after the Transzoic epoch (10th passage of the star). All the water of the Earth would average 6,000 feet today if it were to cover all the Earth's surface evenly.

The graph indicates the depth of water on the other planets during the Silurian Age as follows: Mercury-400', Venus-1800', Mars-5000', Jupiter-8700', Saturn-10,300', Uranus-9,700', Neptune-7,800', Pluto-6,000'. This graph is based upon the fact that the ratio of the lightest elements increases vastly as each planet's distance from the Sun increases.

two planets were placed in a unique position which allowed them to acquire satellites and/or rings revolving in diverse directions. They were also accelerated by the star which allowed them to cross its wake, therefore, placing them in position to intercept debris racing after the star.

Although Mercury and Venus were between the Sun and the star as it entered the solar system, by the time it reached their sector, the planets had moved around behind the Sun, which protected them from serious damage; whereas, the other planets, not moving as swiftly, were caught. Had the Earth and Mars been in a different position in their orbits as the star arrived, they too might have been safely behind the Sun during the passage of the star, and thus escaped damage.

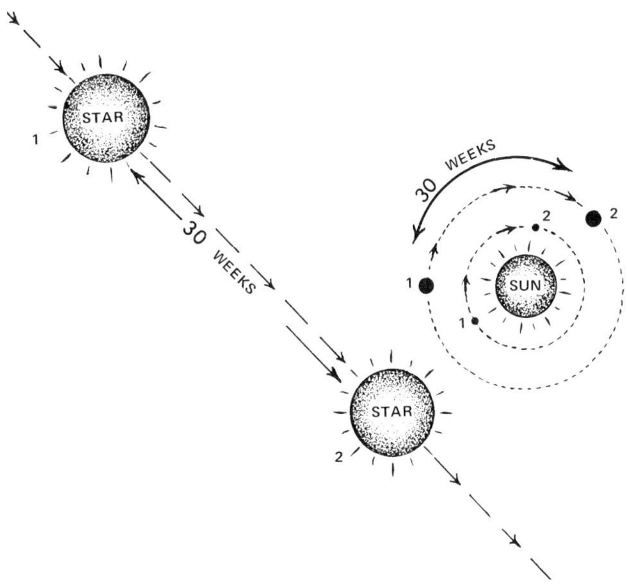

Fig. 52. The orbital positions of Mercury and Venus during the passage of the star.

Uranus was also to the right of the passing star, but it was above it. Uranus was tilted 82 degrees as the star passed far below, yet it lost most of its surface to the strong gravitational pull of the star, leaving Uranus with not only a changed rotation but with a slowed rotation. The star drew Uranus downward in an arc causing a 92 degree inversion of its poles so that the planet's north pole now points almost directly to the Sun (Fig. 53). The planet's five moons revolve on a perpendicular plane to the planetary orbit. *The various orbits and angles of all satellites and Asteroids are the main keys to the proof of a stellar transit through the solar system—and its path.*

The planet Neptune was completely overturned by the star as its satellites were pulled from its surface. The star left Neptune with a 32 degree tilt to its axis and its two satellites moving retrograde to the axial rotation of the planet.

Of all the planets, Pluto suffered the most. The stellar body passed too close — a scant few hundred thousand miles below it. In passing, the attraction of the star completely stripped away the outer crust and all the other outer layers from around Pluto's core.

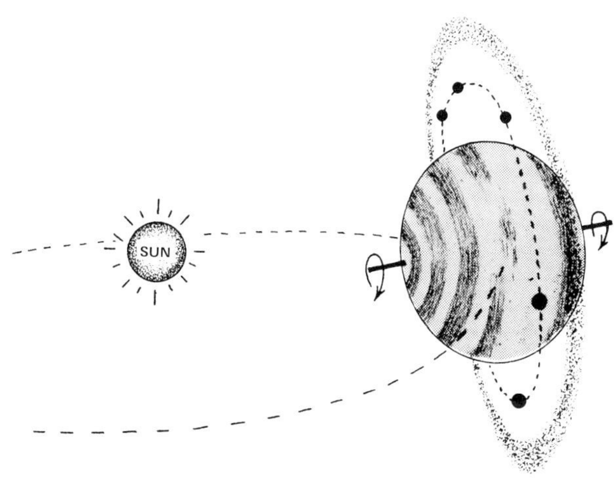

Fig. 53. Uranus and its accompanying satellites and rings have an axis tilt of 82°.

Due to the difference in densities of its crust and core, the crust departed swiftly from around the core. The more dense core was slow to accelerate. The largest crustal portions were absorbed by the star's gravitational field. Originally, Pluto was much larger than Mars, but upon being stripped of all its crust and sub-crustal layers its weight was greatly reduced. It streaked after the star for nearly 3,250,000,000 miles before being left stranded out in space.

Though Pluto was far out in space and only one-fifth its former size, the Sun's gravitational attraction was effective enough to pull the planet into its present orbit. Now only a barren core, Pluto revolves around the Sun 3,680,000,000 miles out in space. When viewed through a telescope, Pluto displays a bright metallic glimmer. This is due to a thin layer of metal remaining over the surface of the core. The eccentricity of its orbit is due to its fast flight in space in its attempt to catch up with the star. This gave it its very high angle and elongated orbit. The eccentricity is so great that Pluto even moves within the orbit of Neptune at times. This angle of Pluto's orbit is additional evidence that points to the direction the star was moving as it passed through our solar system. In addition, it aids us in locating this star now far ahead of us in space.

The Sun and the planets are wonderfully and ingeniously designed. The Earth's complex structure and its various types of layers allow it to continue to function, reshape itself, and maintain its equilibrium in time of crisis or catastrophe.

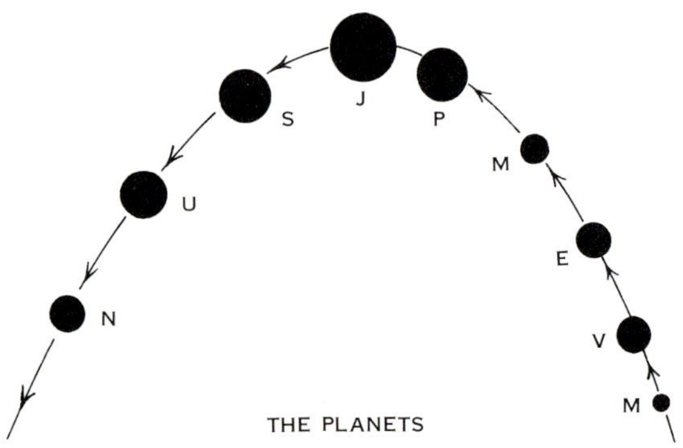

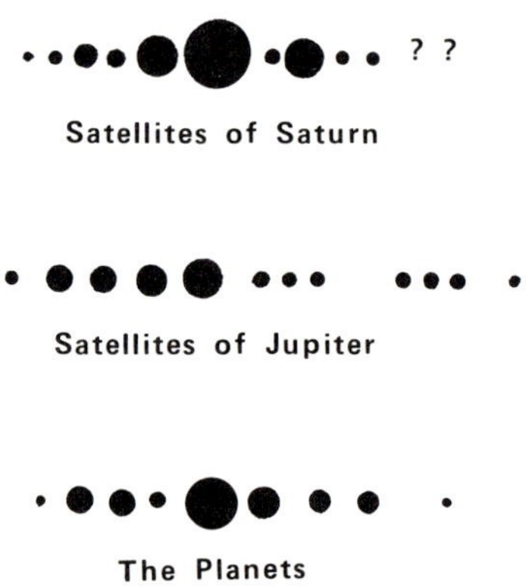

Fig. 54 shows the range of sizes of the planets and Fig. 55 shows the range of satellites that string out from Jupiter, from the smallest to the largest. This indicates only one mode of creation — gravitational force increasing to a peak from minimum to maximum force then back to minimum.

The Birth of the Moon

Fig. 56.

THE BIRTH OF THE MOON

Chapter 4

THE BIRTH OF THE MOON

As the speeding star approached the Earth's sector of the solar system, the inversion of the Earth's poles began. The gravitational pull of the star drew the Earth down and out of orbit in a gradual arc. The pull was steady but with ever increasing intensity (Fig.60).

The star began to pull away portions of the Earth's atmosphere and some surface water. The rising water and air had the appearance of huge drops falling up instead of down. As this air and water reached the outer normal limits of the atmosphere it began to expand rapidly into great cloud-like vapors, then into thin wisps of almost nothingness in the near vacuum of outer space.

On Earth, water gathered and rose into great mountainous tides and began to move slowly over the surface of the ground from east to west. At this time, remember, the Earth was covered from pole to pole with low land masses and shallow seas (200 to 300 feet). There were no mountains, continents, or deep oceans to obstruct the movement of surface water.

As the attraction increased so did the fury of the flood. Water piled up around the equator many miles deep, tapering off on each side near the fifteenth latitude north and forty-fifth latitude south. This great rushing flood smashed trees and all land life in its path. The earth was experiencing great surface destruction. The rampaging seas were filled with mud from the land and sea floors as well as land life, marine life, trees and plants. Anything that could be picked up was swept along with the water. The tempo of life had reached its zenith at this time, plant, animal and marine life were at their pinnacle. The biosphere of Earth had nearly reached the saturation point, so one can imagine the density and variety of life and material that filled the rushing waters.

The great flood of water gradually changed latitude as the Earth inverted, with the result that after passing the 70th latitude the rotation of the Earth caused the water to reverse its direction of flow and move towards the north pole which was beginning to point toward the star. As the water reached the north pole (now pointing toward the star), because of the rotation of the Earth, it swirled in a whirlpool motion just prior to its exodus into space.†

Fig. 57. Surging waters on the Earth.

† If one views the Moon today, one can still see, on many areas of its surface, the remains of the flow lines of these great volumes of water as they rushed over the surface prior to the departure of the crustal section from the Earth.

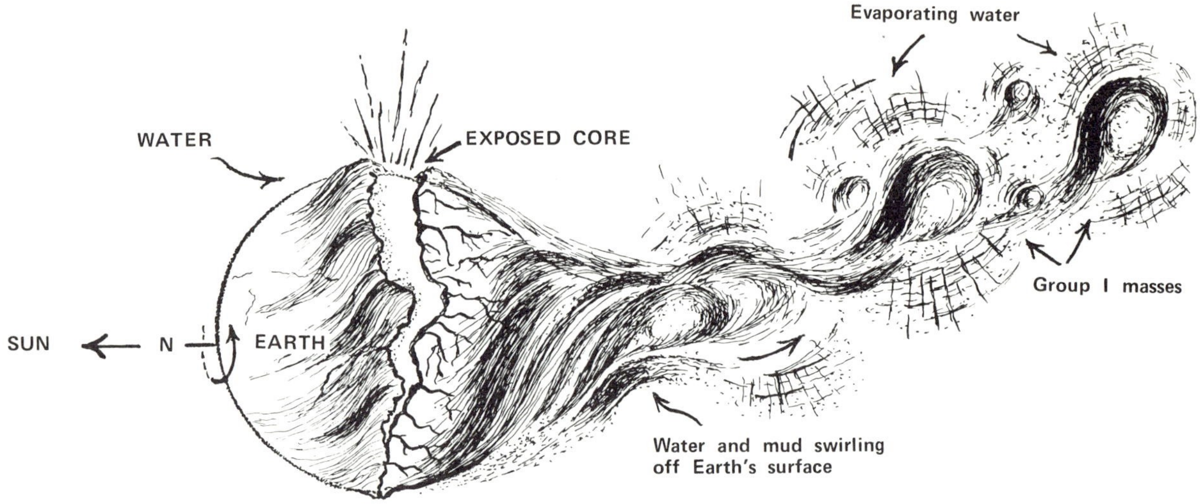

Fig. 58. The atmosphere and waters of Earth departing forever. The rings of Saturn contain some of the atmosphere and much of the waters of Earth. Venus and Jupiter acquired the balance of Earth's atmosphere.

This great tidal flood, full of material, was pulled from the Earth in a chain of masses before the actual separation of the crust. They raced after the star in great writhing globules. Though outer portions of the liquids in the masses of water, salts and Earth debris evaporated in the vacuum of space, the main masses remained intact because of their density and size, which gave them a gravitational value.

Slowly the present South pole (which was then on "top" – the original "North Pole.") turned over and down in a great arc until it was pointing away from the Sun and pointing directly toward the star as the star crossed the plane of the planetary orbits (Fig. 60).

To obtain greater detail of the activity that occurred prior to the moment the star arrived in opposition to the Earth, we will go back a few weeks in time when the gravity of Mazzaroth began to seriously affect the Earth –

For weeks, while the Earth turned on its axis, the crust began to roll and dance crazily. Subterranean shocks increased in intensity. Air screamed over the surface of the ground as it rushed toward the equator from north to south. Great temblors began to jar the surface of the Earth as though giant pile drivers were pounding from below.*

Then, as the star drew closer to the Earth, massive cracks opened in the crust, particularly in

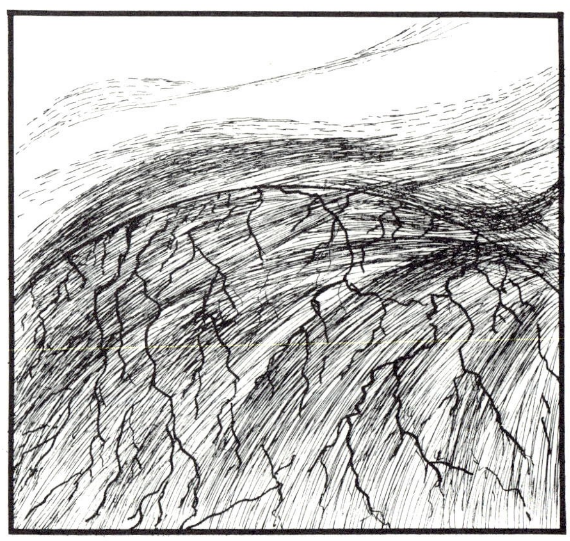

Fig. 59. Air screamed over the surface of the ground.

* Due to gigantic surges of the metallic layers.

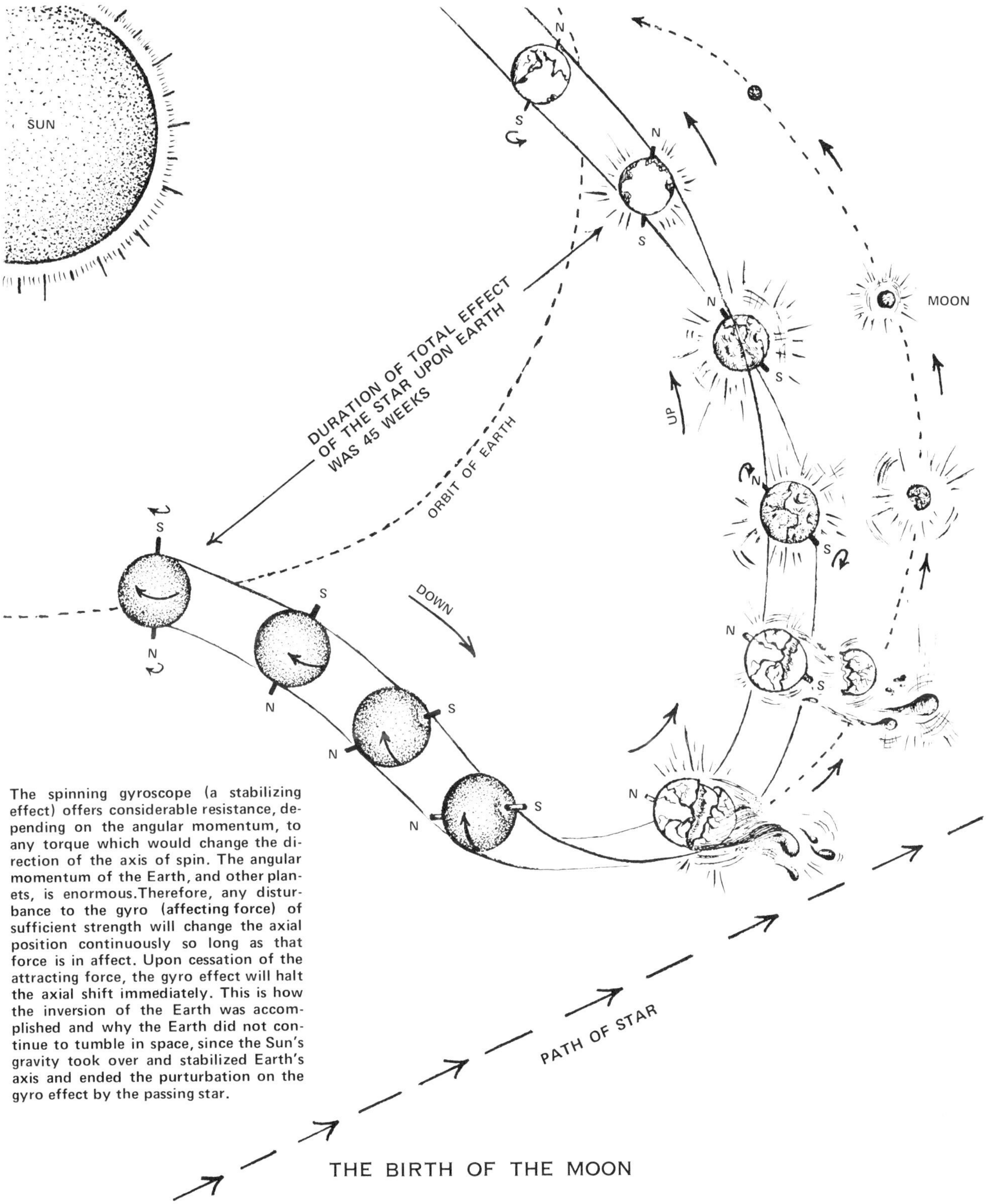

Fig. 60. The inversion of the Earth.

Fig. 61. The birth of sand.

Fig. 62. Sand dunes in Death Valley, California (from a great fault nearby). These dunes are almost identical to Great Sands National Park in Saguache County, Colorado.

the area opposite the approaching star. Huge sections of crust broke off and sloughed down into them; water in great volumes began cascading into these yawning, flame-belching chasms. As the water poured down, it struck rising molten rock and metals. Gigantic and terrible explosions resulted.

All over the Earth, stones, boulders, small rock fragments and fine dust and sand began to blow out from hundreds of thousands of massive fractures – *here was the birthdate of gravel!* Billions of tons of gravel shot into the sky in a steady stream and peppered down over the Earth's surface as huge sections of crust surged back, then slammed together continuously squirting great volumes of liquid rock into the sky at high speeds.

Sand came down like rain as rock from great depths came up to the surface as a gas which condensed the moment it reached the cooler outer surface of the Earth. It was from every fracture of the crust that more sand was produced as water struck hot rising, liquid quartz, granite, basalt and all the other pyroclastic materials.

Fig. 63. Gravel lying on the surface of the ground near Joseph City, Arizona. All this gravel was slightly plastic upon landing. Note impact points (white spots).

Fig. 64. The above gravel arrived out of a nearby fault and landed on the surface of the rock strata adjacent to the fault.

Fig. 65 and 66. Sand that originated from a nearby fault. Sand dunes such as these can be found worldwide.

U.S. Department of the Interior - National Park Service photo.

The Earth was caught between the pull of gravity of the two suns, each of which battled for possession. Something had to give. At that moment one half of the Earth's crust – the southern half –* was forcibly removed. It was literally slipped away like the skin from a ripe peach. At that same moment, great volumes of metals shot out of the Earth into the departing mass.

The thin section of the Earth's outer crust had the appearance of a huge bowl floating upside down in space. From the bottom of the cavity (64 miles deep) just created in the Earth came a brilliant, eerie glow that penetrated and lighted up the dust filled sky with a blinding light. The quartz, basalt, feldspar and orthoclase liquid rock layers lay exposed for the first time since the earth cooled billions of years earlier. But the layers were exposed only briefly, crustal materials and liquid rock and metals began rushing in to fill the great open wound in the Earth.

In a very short time the remaining half of the Earth's crust had broken apart in many places. The numerous sections, floating on a sea of molten metal and rock, moved toward and closed the vast chasm until equilibrium was again established on the Earth. This is how the continents were formed and why they seem to fit together like pieces of a giant puzzle.

Although the Earth continued to accelerate during the activity, the direction of the star's attraction changed rapidly. At the moment of the crust's removal, the star passed *above* the plane of the planetary orbits. This is the action that caused the inversion of the Earth. As the Earth continued its gradual inversion (Fig. 69), it began to curve up in an attempt to follow the star. At the moment the north pole passed vertical by 23½ degrees, the Sun's attraction overcame that of the star. The Earth's inversion was arrested and it was drawn back into orbit.

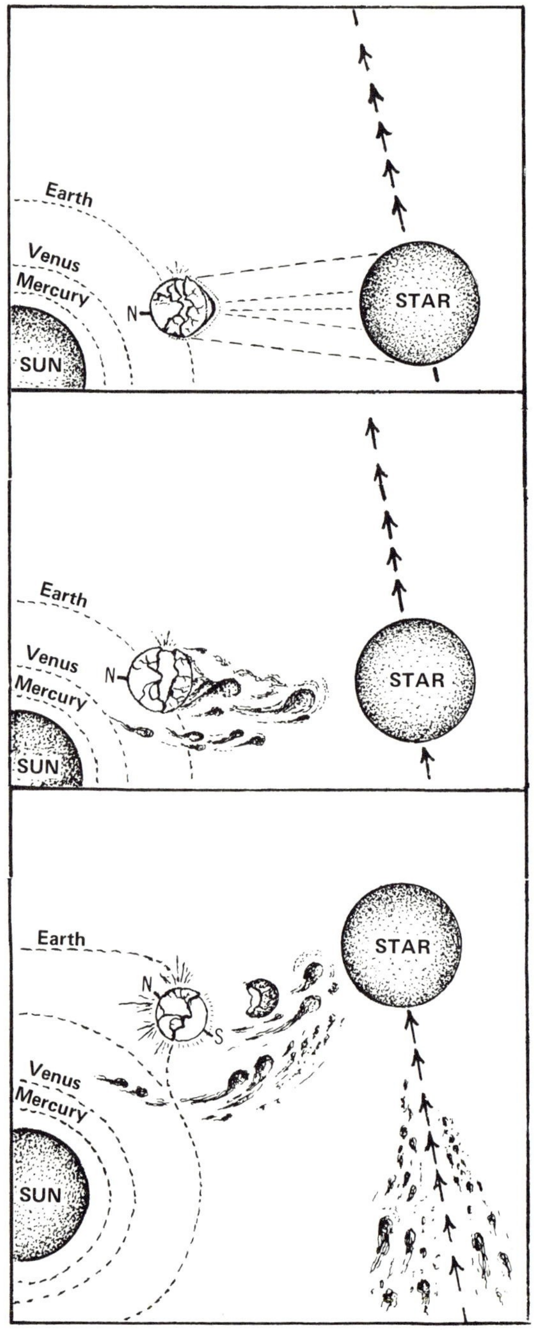

Fig. 67, 68, 69. The Moon's birth sequence.

* Actually the Earth's original northern half.

At first, it would seem that the Earth, having lost 5,290,790,400 cubic miles of mass would have assumed a smaller orbit around the Sun. But the Earth retained the Moon as a satellite, gained the waters of Mars and some from Venus, so the total mass remained about the same; therefore, the orbital position also remained about the same.

In addition to one half of the Earth's crust, the gravity of the star pulled off a few smaller sections ranging in size from small particles to sec-

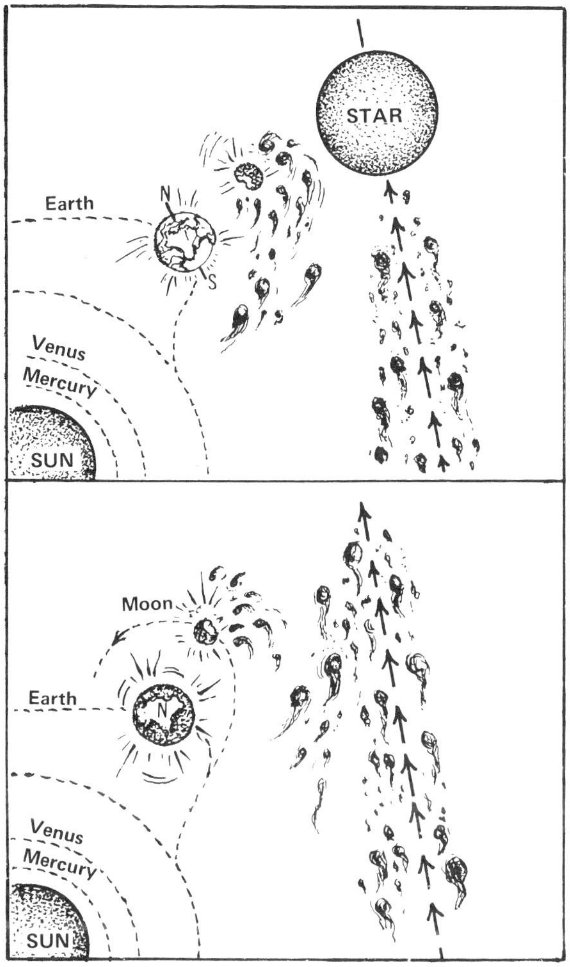

Fig. 70 and 71.

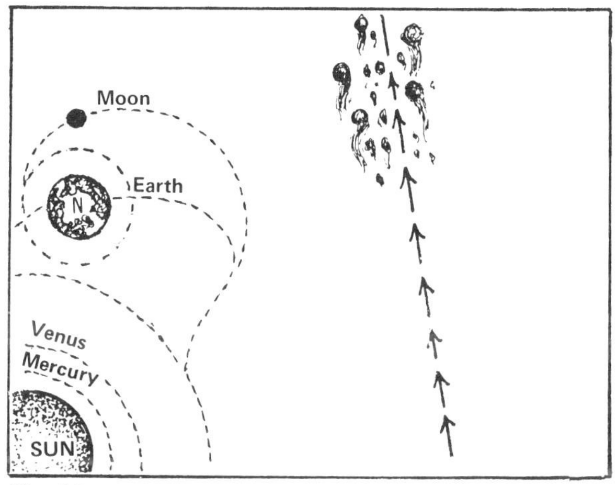

Fig. 72.

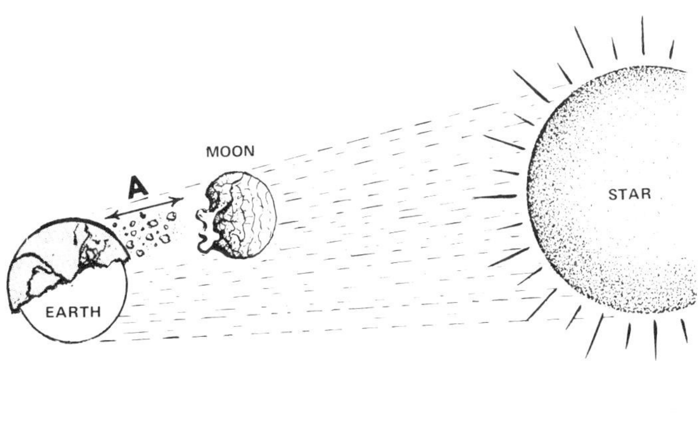

Fig. 73. Some boulders at point "A" left the Earth a short distance, then gently drifted back in the nearly neutral gravity field. They now stand as isolated out-of-place "mysteries" (Fig. 74.).

Fig. 74. Erratics such as these were said to have been dropped by retreating glacers.

tions 160 miles in diameter. Many many of these fragments were destined to become meteors. Some of the largest masses would become asteroids or comets.

The huge bowl-like section of the crust accelerated until it reached a maximum speed of 2,280 miles per hour. By the time it had attained this speed, the stellar body to which it was being drawn had passed beyond its field of attraction, and its effect upon the Earth and the crustal section was diminishing rapidly.

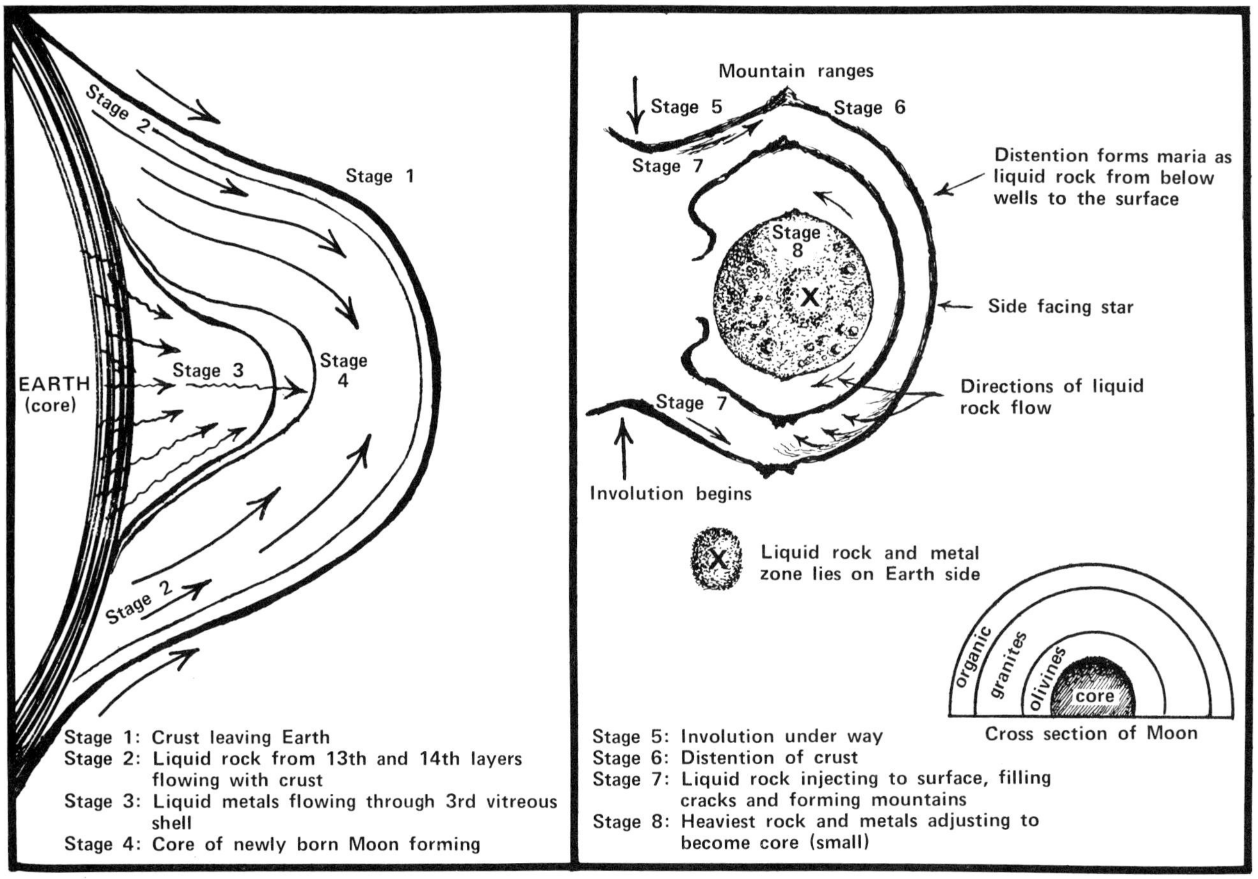

Fig. 75. Stages of the birth of the Moon.

Immediately after being pulled from the Earth the bowl-like section of the crust began to collapse upon itself in an attempt to form a sphere. The fringes of the crust began involuting until they were drowned by hot liquid magma which flowed into the center from the inner side of the crust (Fig. 75).* The gravitational force within the crustal material was sufficient to pull the outer edges toward the involuted center. There was no true core material present around which the crust could move, but liquid and plastic rock layers were fluid and flexible enough for the harder and more rigid materials to slide and move upon. All layers during this activity became a completely disorganized, jumbled conglomeration, obliterating nearly all traces of an organized body.

* Seismic units operating on the Moon have found a discontinuity in the physical properties of the rock at a depth of about 60 kilometers (36 miles). This confirms the layering sequence of Earth (See Fig. 32, page 42). Below this is rock of higher density.

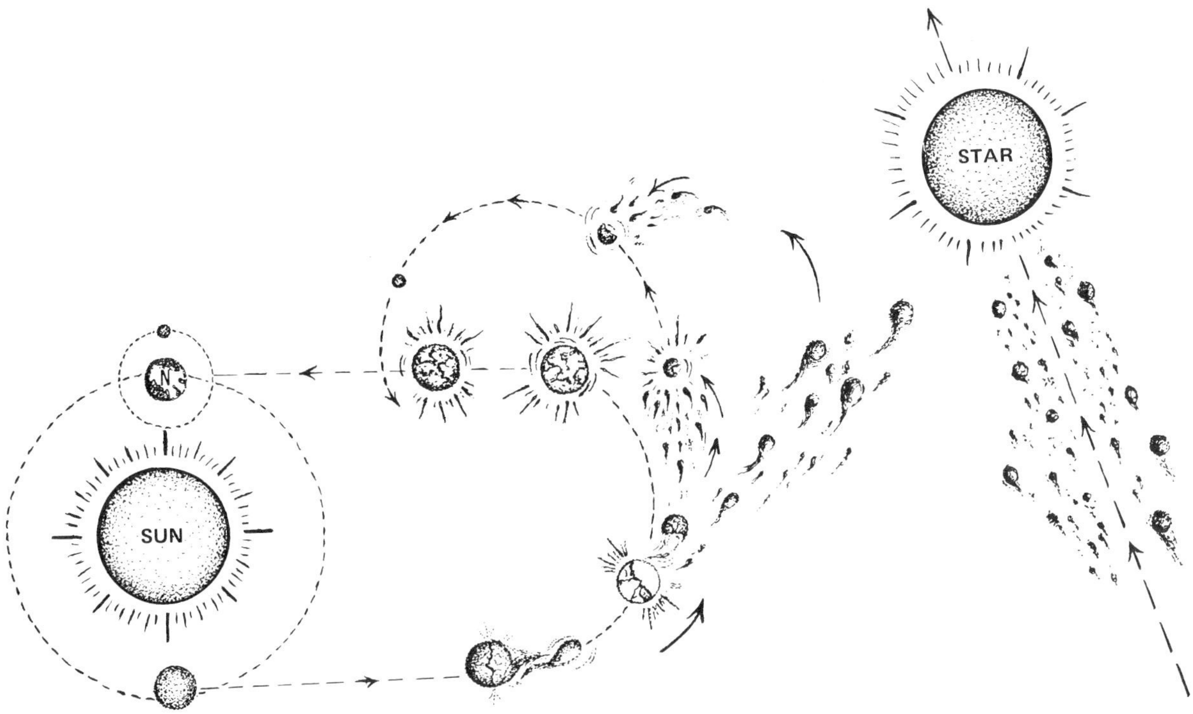

Fig. 76. A "moving picture" of the moon's creation.

When the collapsing section reached a point approximately 180,000 miles out in space, the gravitational attraction of the earth, aided by the Sun, began to overcome that of the passing star. The Earth began to pull the section out of its course, causing it to gradually curve away from its direction of movement while maintaining its convex face toward the Earth.

Upon reaching a distance of 240,000 miles from Earth, the crustal section was pulled into a permanent orbit around its parent. Our satellite, the Moon, was born.

Crustal distention of the newly created Moon was in progress as the first of the great globules of water, the liquid saturated masses of debris, and other fragmented material from the catastrophe arrived to pit and crater the still soft surface.

During the initial distention every fracture in the solid layers was instantly injected with liquid rock the moment a fracture appeared. Hot liquid rock flowed from every deep fissure and leveled out evenly, filling the broad areas of distention, and cooled as it arrived. One has only to look at the interesting features called "ridges" on the floors of Mare Serenitatis, Mare Imbrium, for example to clearly see the evidence of this activity. Meanwhile, the crust continued to distend over the newly forming core.

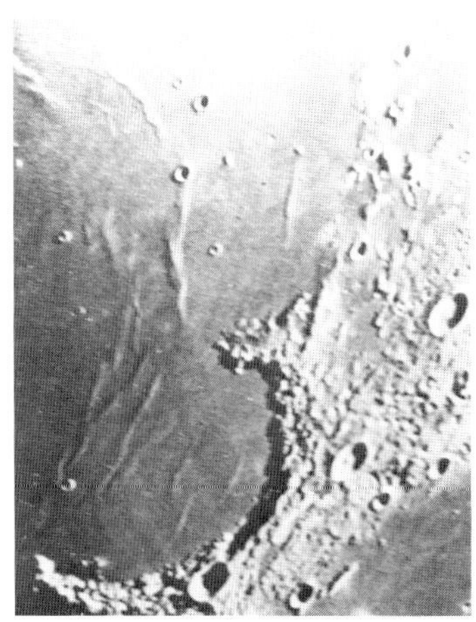

Fig. 77. Ridges on the Moon were created by the crust closing in on the edge of the maria containing the mascons of metal.

However, since the liquid quartz-granite layer (13th and 14th – See Fig. 32) had injected the bulk of its mass into the initial distention, the more viscous rock layers had to complete the leveling and filling as the balance of the crust closed up the gaping void that had faced the Earth's inner layers, leaving the bulk of its weight on the side facing the Earth. Consequently, the Moon remained out of balance with her center of gravity toward the face she always presents to the Earth. This is the basis for her "captured rotation." The other satellites in the solar system also always present the same face to their parents – simply because the same process of birth and adjustment occurred on them.

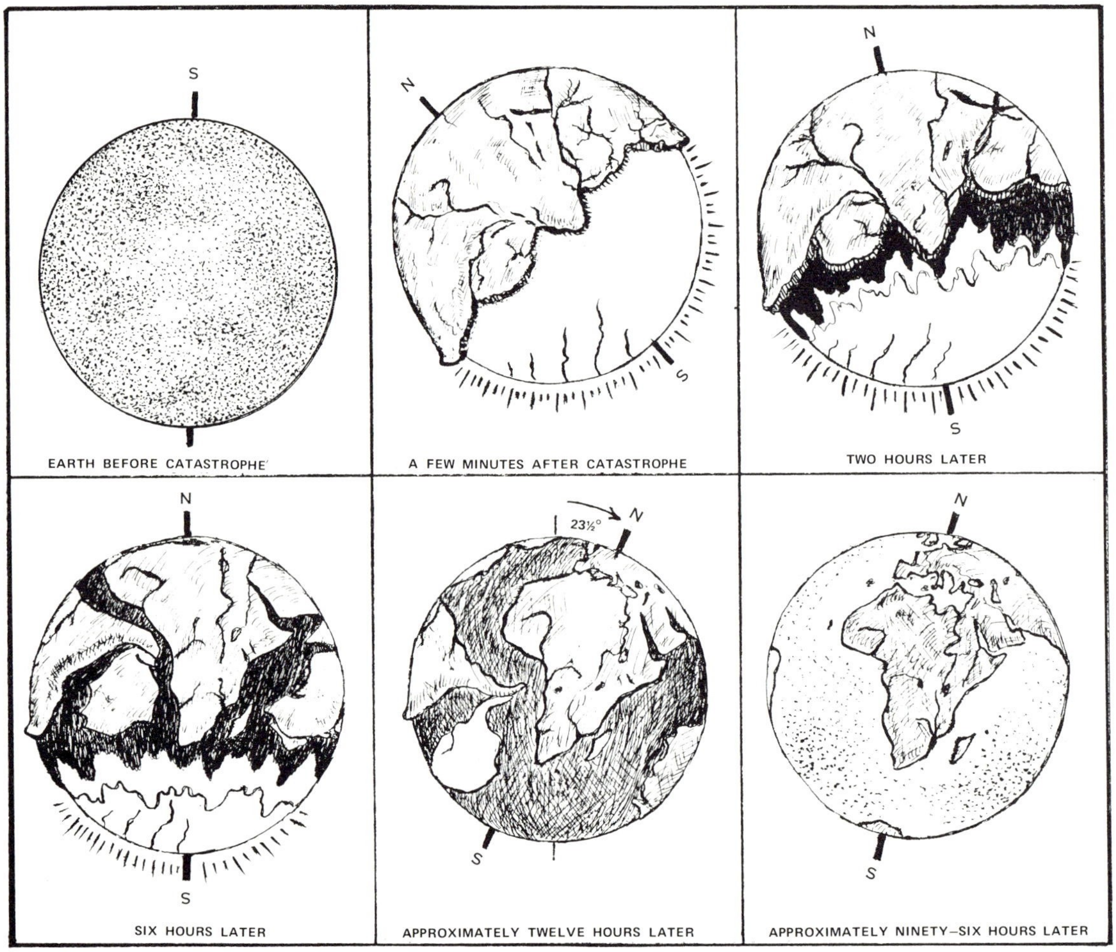

Fig. 78 The Earth's crust adjustment after the Moon's departure.

It must be noted here that just as the crust was parting from the Earth, great fractures in the First Vitreous Interbasement, Second Vitreous Interbasement and Third Vitreous Interbasement layers of the Earth's crust (See Fig. 32) opened for a few moments allowing the extrusion of a great volume of metals into the departing cavity of the extracted crust. These metals originated from the Semi-fluid, Second, Third and Fourth Intermetallic layers of the Earth. Due to the enormous attraction of the star, the metals surged and pounded against the solid layers until they fractured. This then allowed these metallic fluids to flow in great volumes at high speeds out of their layers into the departing Moon. It was these metals that really created the "heavyside" of the Moon and prevented rotation.

Orbiting vehicles of our space program, using gravity detecting equipment found five anomalies, called "Mascons" (massive concentrations). These anomalies are all found within the marias of the Moon and the marias are filled with these deep seated metals, samples of which were brought back by the Apollo crews.

Some of the rocks the Apollo crews brought back were classified as "Kreep norite" which is found only in the marias. This material consists of uranium, thorium and potassium. The origin of this type of rock is one of the most puzzling questions confronting other scientists today. Yet it is not a problem when it is understood that this material originated from deep within the Earth at the moment of parturition under the influence of enormous forces.

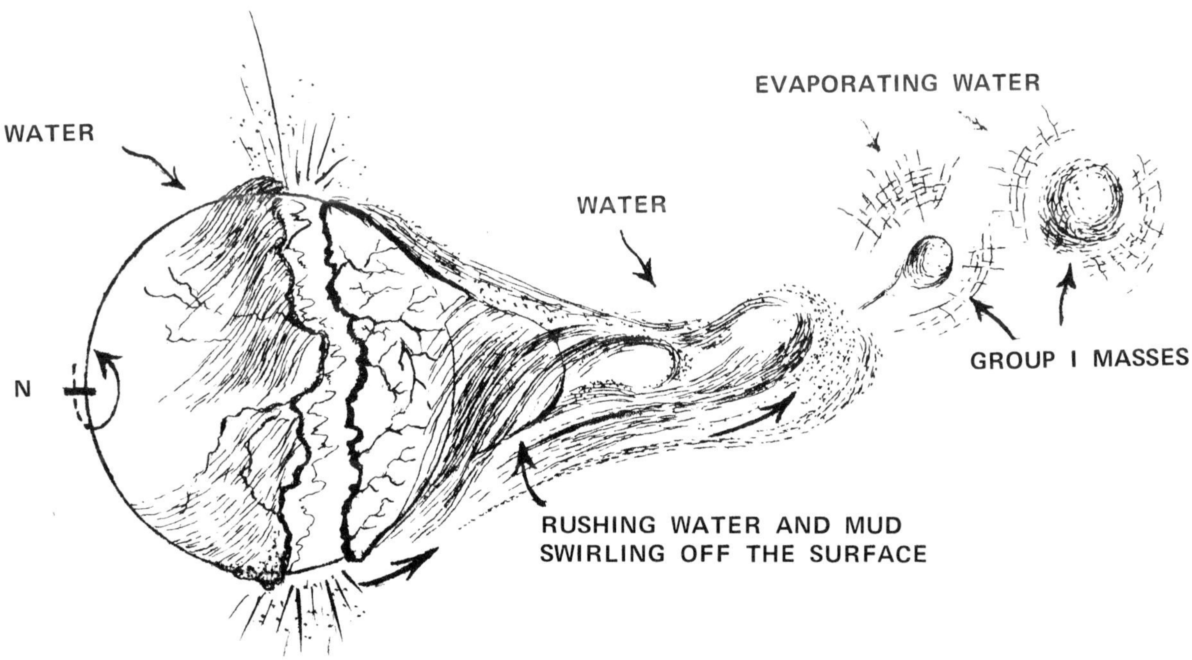

Fig. 79. Early stage of the birth of the Moon.

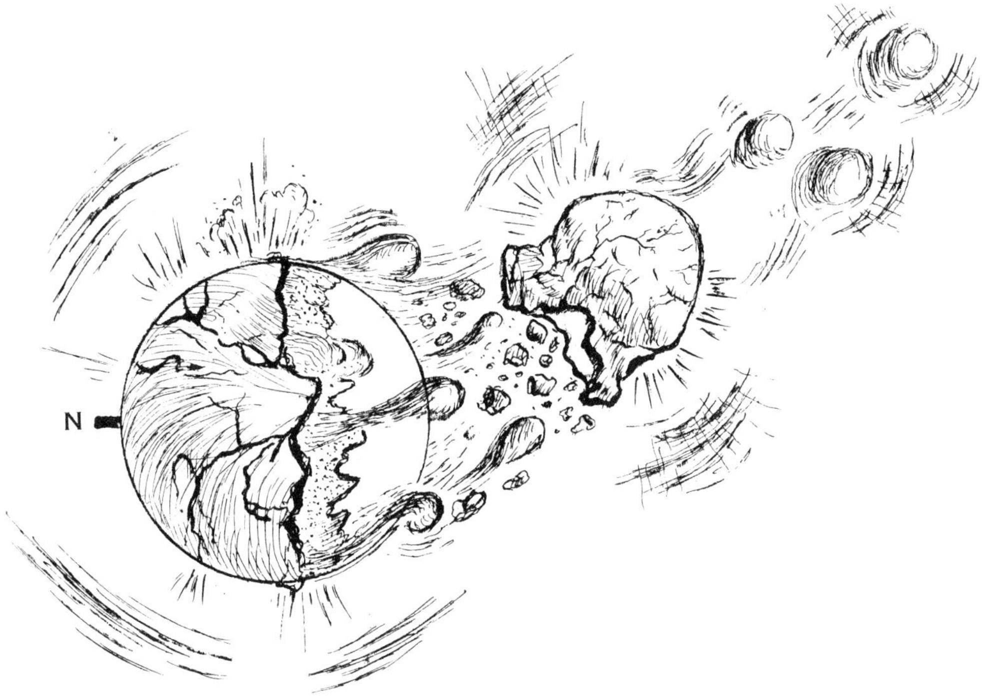

Fig. 80 and 81. Later stages of the birth of the Moon.

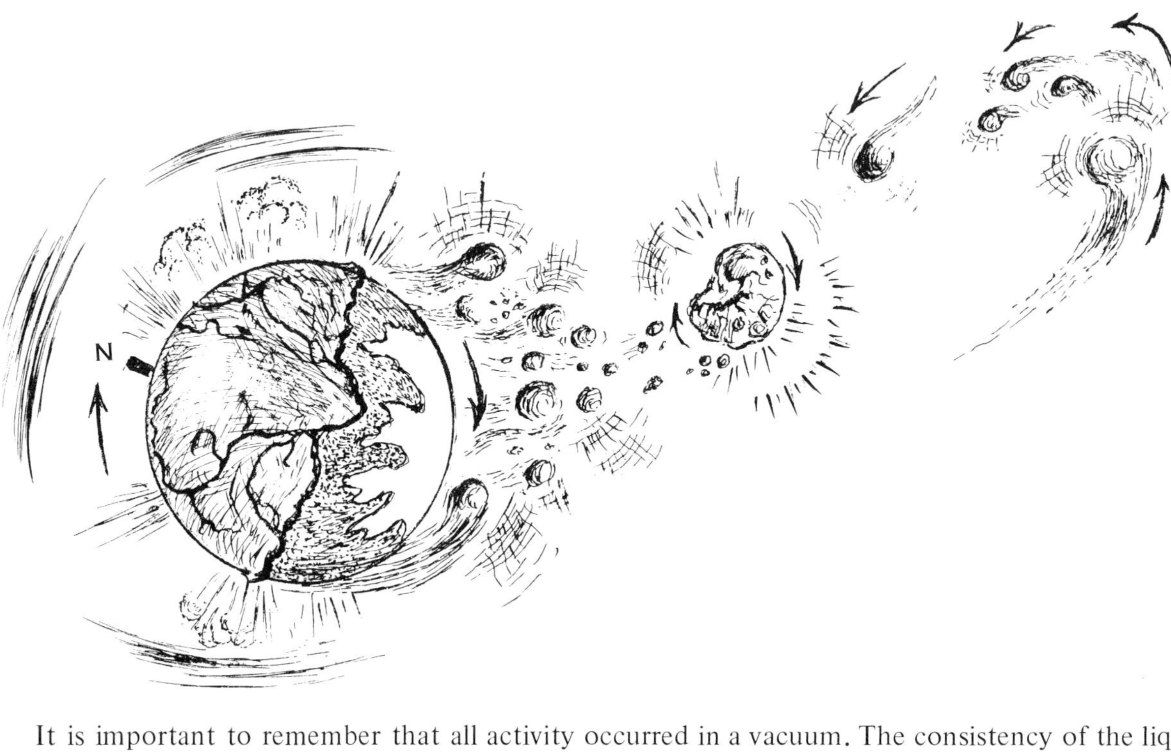

It is important to remember that all activity occurred in a vacuum. The consistency of the liquid masses varied greatly. Some departed from the Earth containing only natural salts, while some were largely mud and surface debris.

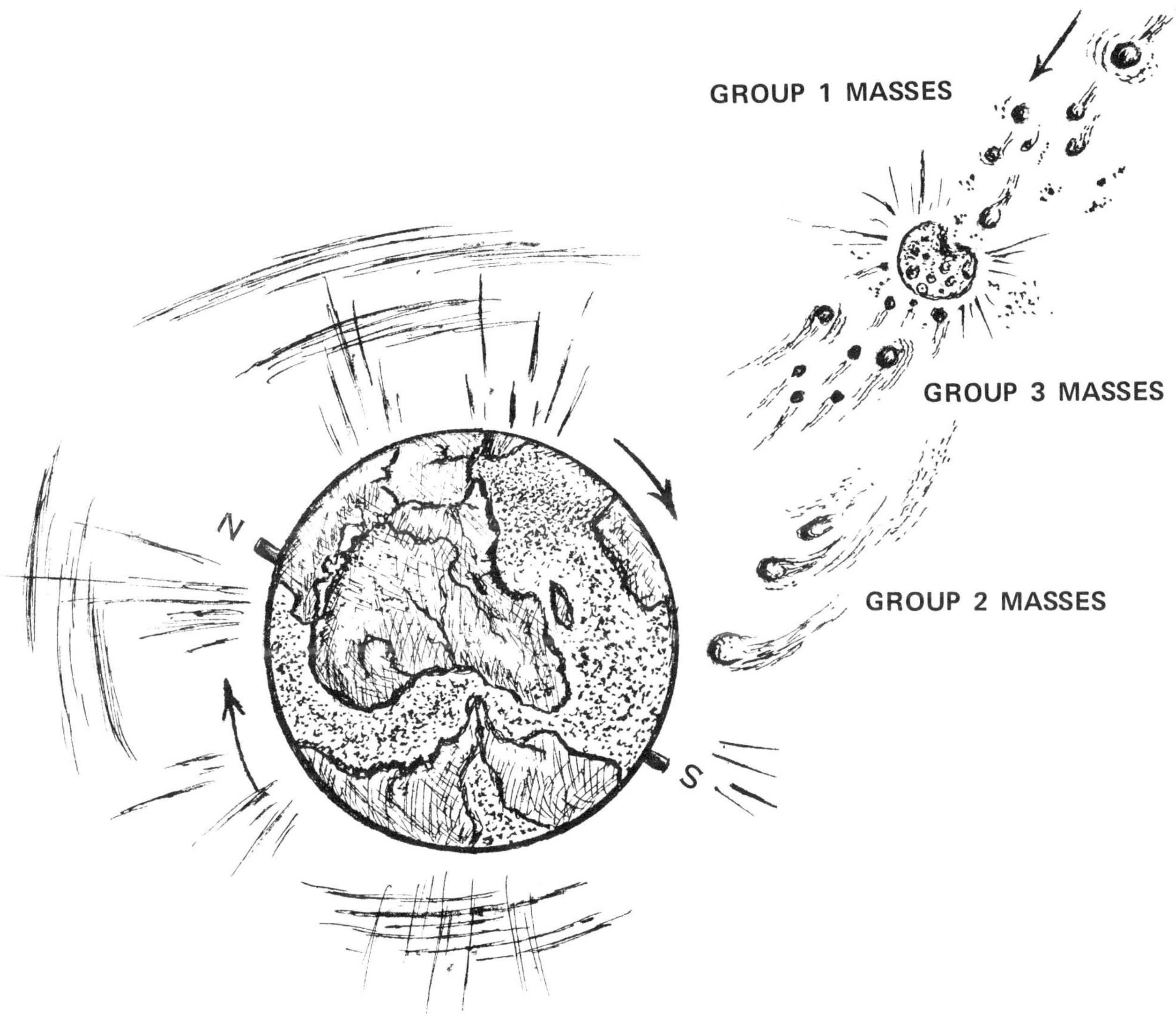

Fig. 82.

Fig. 83. NASA

THE MOON

The Craters of the Moon

Fig. 84. HALE OBSERVATORIES

THE MOON

Chapter 5

THE CRATERS OF THE MOON

Much of the Cretaceous upper layers were flushed off the Earth's surface and were deposited on the Moon since hundreds of feet of that material had not as yet hardened into rock. (Forming of rock strata occurred by gradual hardening under the constant covering of water and required several hundred thousand years.)

On exposure to the atmosphere as a result of the catastrophe, the muds and soft rock hardened quickly on Earth (within ten days) into hard rock. The muds which landed on the Moon and those that were the surface of the Moon at the time of the catastrophe never hardened but became like a powder in the absence of an atmosphere.

The craters of the Moon were formed by three classes of incoming material, each class arrived at a different stage of the moon-shaping activity. These materials were in the form of numerous globular masses from fluid to plastic consistency. Some were water, or different surface materials mixed with water. While a small number were molten quartz-granite rock. The size of these globes ranged from the size of raindrops to great blobs hundreds of miles in diameter.

Group I and Group II masses, defined, are those globes of mud, ocean water, marine debris, plants, trees and animal life that were pulled off the Earth. Group I is that group which preceded the removal of the Moon. Group II is that material which left after the removal of the Moon. Group II had a higher precentage of surface material than Group I, which was mainly water. Group III consisted of small amounts of liquid granite which followed the Moon and peppered its surface almost the moment involution ended.

The following is the sequence of events of impacting material on the surface of the Moon just after its removal from the Earth and while it was in the process of mechanical readjustment into its spherical shape.

The first material to impact was made up of anything that could be swept off the face of the Earth – water, soil (mud), soft cretaceous rock, trees, plants, animals and marine life.* These materials struck the still wet, mud-coated crust of the Moon as distention of the surface began.

While distention of the Moon's crust was in progress and the liquid rock layers were welling to the surface through the distended fissures to flood out and become the floors of the marias (dark level areas), great soggy blobs of mud and surface debris smashed into these still soft marias. Traces of these impacts are detected as "ghost craters" such as are seen on the floors of Mare Nubium. The

*** If this organic material had remained on the Earth instead of being swept away, all soils would be extremely rich, chemicals would abound and oil fields would contain much more than their present potential.**

numerous "bays" bordering the many maria, such as: Sinus Iridum, Sinus Beaumont, Doppelmayer and Hippalus were formed by material that impacted while the maria were still plastic.

The Moon's largest craters do not have center pinnacles on their floors simply because the material that formed these craters was quite fluid and low in solids. This also resulted in greater diameter craters with low rim elevations.

The consistency of the blobs and globules of material varied greatly. Some that left the Earth were only sea water full of its natural salts (Group I), while others were masses of very wet debris-laden mud (Group II).

The larger maria, such as Mare Imbrium, is an excellent example of massive Group III arrivals. The inner part of this basin sank to a level below the surrounding terrain. This sinking was due to crustal down-warping which in turn caused the Caucasus, Apennines and Carpathian Mountains to be extruded to the surface. The "rings" that surround the mare and which have the appearance of fractures, are due to waves of igneous rock that welled up during crustal distention. These waves of liquid rock solidified before they could level out with the surrounding surface.

Fig. 85. Mare Imbrium, Mare Serenitatis, and Mare Tranquillitatis. **U.S. Government Printing Office.**

Fig. 86. Mare Crisium NASA

LUNAR "SEA"

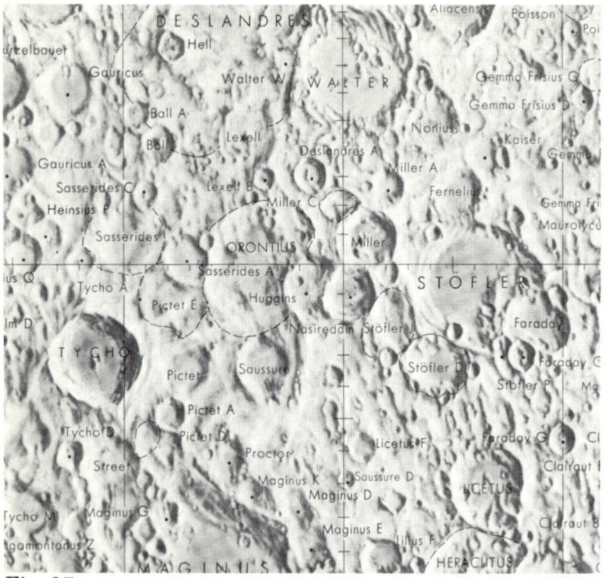

Fig. 87. **U.S. Government Printing Office**

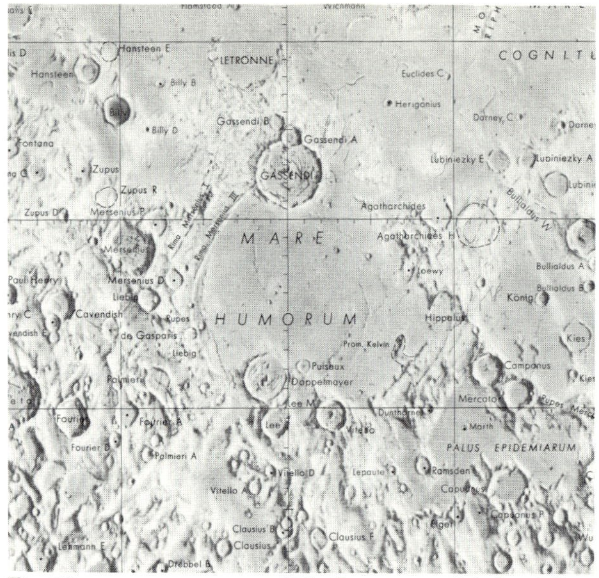

Fig. 88. **U.S. Government Printing Office**

There are some important features to keep in mind regarding the material that left the Earth. As these great streaking globate masses rushed through space, their surface waters converted to vapor.* This resulted in the large quantities of mud being crusted over on the outside but still wet and plastic within. These "mud balls" rushed toward the newly born Moon in great numbers. One has only to glance at the Earth side of the Moon and observe the number of craters large and small to get an idea of the number and varying sizes of the individual "mud balls" that landed on the Moon. However, the materials that impacted the Moon's surface were only about one-fourth the diameter of the craters they formed.

At this time the second group of materials to leave the Earth began to rain down on the Moon's still-plastic surface in great numbers. They impacted simultaneously on the southeastern side, then tapered off to numerous smaller stragglers in the northwestern quadrant of the Moon. The photograph on page 80 shows the evidence of this activity.

Though the impacts were frequent and in great numbers none of them were violent since at that time the Moon was in a rather neutral zone between the passing star and the Earth. Being in this unique position the gravity field in the area of the Moon was quite weak. This is why so many craters have large diameters but are of shallow depth. The earlier generated speed of the impacting masses was the only force at play.

The rays around the craters Copernicus and Kepler appear scattered in an erratic manner, those around such craters as Proculus and Thales are along one side, while the rays around Tycho are evenly and widely distributed. These different patterns indicate *different angles* of arrival of the masses that created the craters. In the case of Thales and Proculus the material arrived at a sharp angle causing a splatter of 180 degrees from the crater on one side. While those of Tycho, Copernicus and Kepler were more directly vertical impacts.

*** Where the star passed the Earth at its closest point, the intense radiation of the larger stellar body caused the waters to rapidly evaporate away from each of the various sized masses of water and mud globules that it had pulled off the Earth. Thus, dehydration of these large bodies from Earth was accomplished quickly.**

Fig. 89.

Fig. 90.

Fig. 91.

Fig. 92.

NASA Photographs

Fig. 93.

Fig. 94.

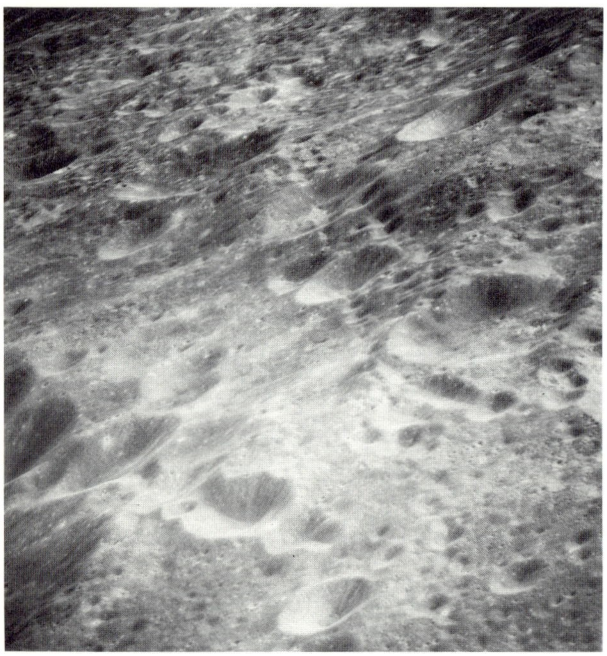

Fig. 95.

NOTE: As you study these or any photographs of the Moon's surface, keep in mind that the material was still moist or liquid when it landed.

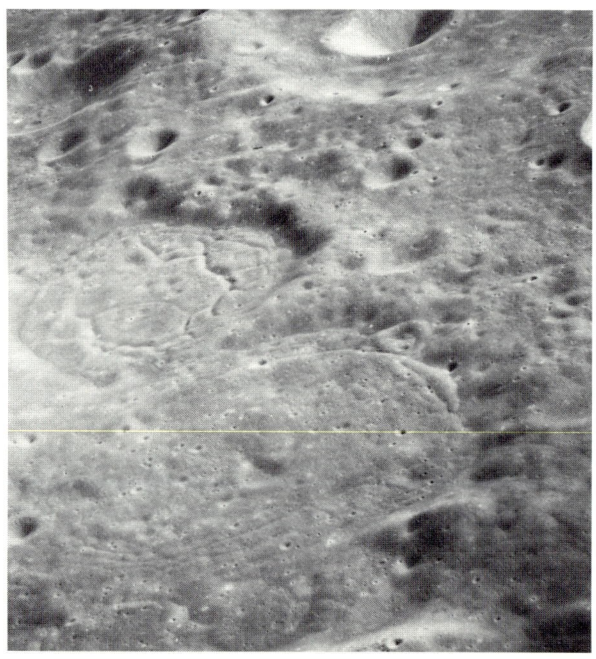

NASA Photographs

Fig. 96.

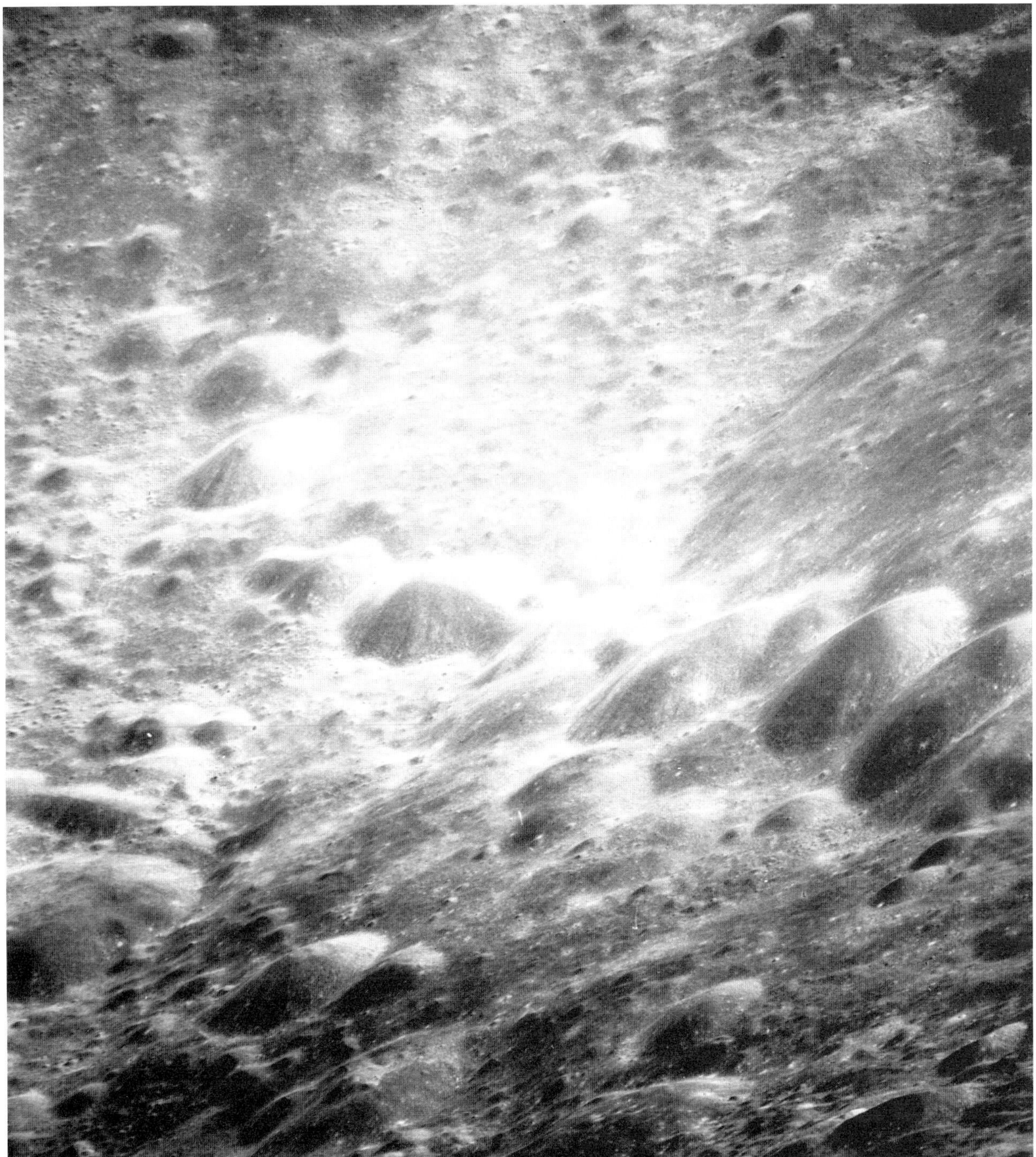

Fig. 97. NASA

NOTE: In some of these photographs, an optical illusion will occur in which the craters will appear to be domes. To correct this, look away (usually down) or turn the picture upside down and the illusion will disappear.

Many craters, when in clusters or in close contact, show distinctive geometric shapes. Excellent examples of these are: Ptolemaeus, Alphonsus, Arizachel, Walter, Regiomontanus, and Purbach. Unusually heavy impact activity can be seen in the region of these craters. The largest craters were created first by the incoming masses.* These craters were defaced almost immediately by smaller blobs of matter, and so on – the smallest craters being created last. This is the reason that, when one crater overlaps another or breaks into it, the smallest crater always defaces the larger in all observed cases.

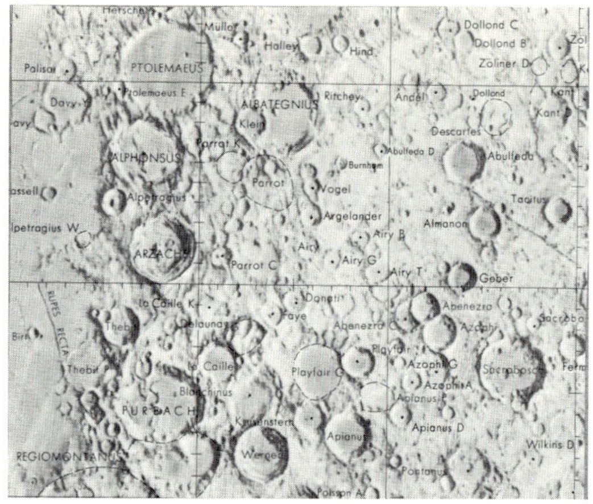

Fig. 98. **U.S. Government Printing Office**

Rarely, as shown in Fig. 99, does a larger impact deface the wall of a smaller adjacent crater. This picture is an excellent example of the draining away of moisture from loose muddy soil that made up the crater rim. After the draining away, the crater wall sloughed in successive stages toward the crater floor.

Fig. 99. **NASA**

As a rule, whenever two craters overlap, the smaller crater is invariably the last to have formed. These could hardly have been formed by meteorites for the explosion caused by an impacting meteorite would have destroyed any pre-existing crater's walls, yet previous walls are undamaged wherever a newer crater overlaps an older one.

When two craters of about the same size are found close together, they are always of similar construction, depth and appearance. They both have central peaks or they do not. Here again, it is obvious that the various sized groups of arriving muds and salts were traveling very close together. Twin craters were created by identical materials which impacted at the same slow rate of speed; whereas, meteorites would not arrive in this manner and certainly not at low speeds. All impacts were made before the surface material became dehydrated or dried into a hard crust.

Fig. 100. **Mount Wilson and Palomar Observatories**

The hexagonal and polygonal shapes of the craters are due to simultaneous arrival of material. As each mass impacted, the shock waves from one crater caused a distortion to the shock wave from an adjacent crater, and so on. Similar materials landed on the same areas, on top of or next to one another, creating similar craters, similar in shape, size

* Larger masses were always pulled away from any planet first with successive smaller ones following.

and detail of construction. Observe the three craters: Walter, Regiomontanus and Purbach, for instance. Most of the craters are round, however, with smoothly symmetrical rims. For only when wet, muddy or plastic masses landed simultaneously did the mutual distortion of each impact shock-wave cause the hexagonal or polygonal shape.

Fig. 101.

Fig. 102. The large crater is Kepler. The smaller symetrical one to the right is Kepler A.

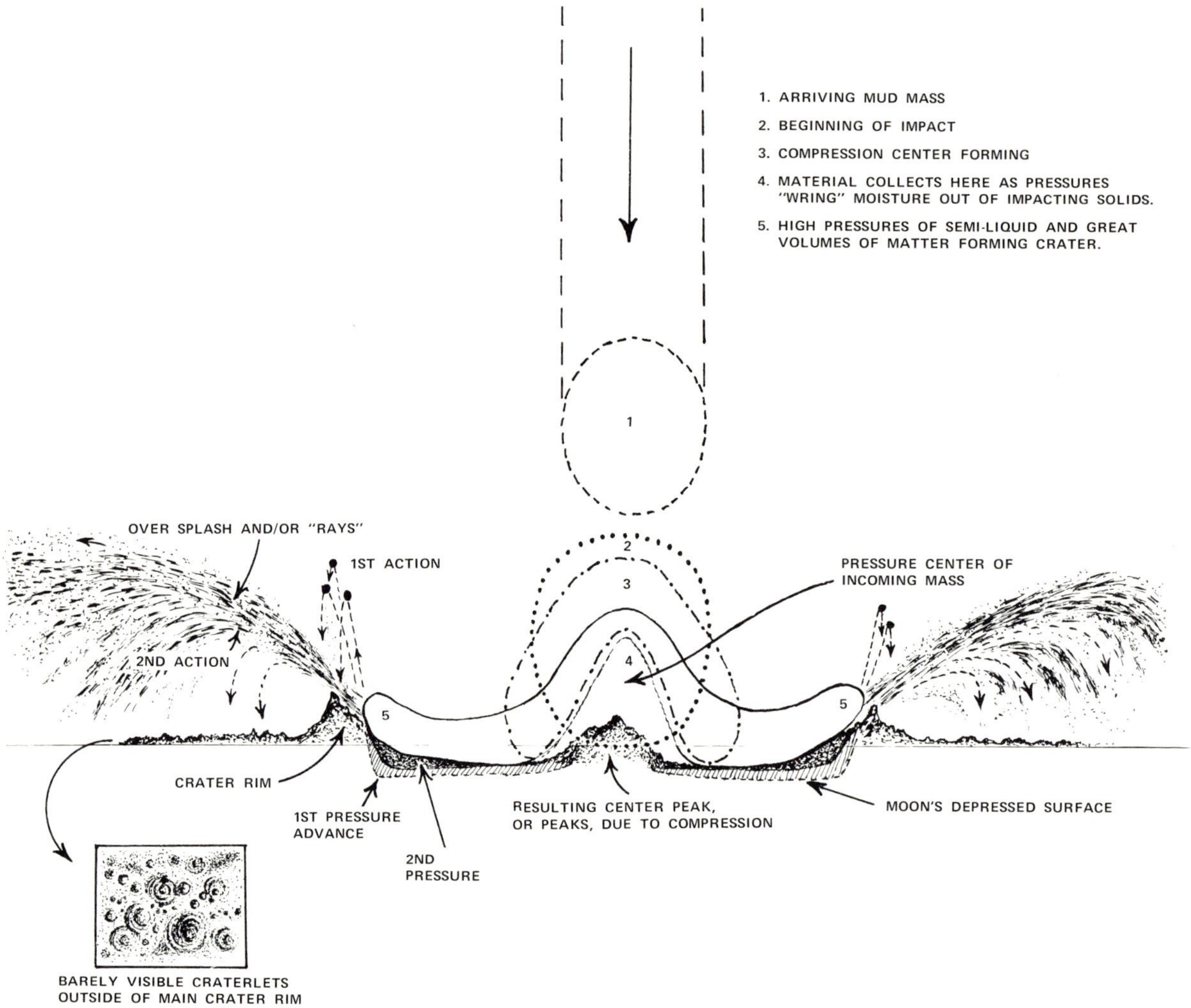

Fig. 103.

FORMATION OF A CRATER

One of the most interesting periods during the sequence of events in the formation of a crater by an incoming mass was the moment of impact (Fig 103). At impact, the surface of the Moon directly under the point of contact was instantly compressed. The compression was so great that further penetration to any appreciable depth by the incoming material was not possible (the relative shallow depth of the crater also indicated low speed). The viscous mass, therefore, spread out horizontally in all directions. This activity created a pressure zone at the perimeter, which at maximum pressure, formed a ridge. The maximum pressure was reached when the greatest portion of the incoming material had impacted. Since the highest pressure point was in the center of the forming crater floor, this point of resistance was left in a compressed state which resulted in peaks of compacted material being left standing in the center of the crater when the water was squeezed out and after activity ceased. This explains the puzzle of the central peaks found in so many craters.

NOTE: **If meteors had caused the craters of the Moon, the craters would be very deep and would lack the familiar ring of small craterlets around the rims. Also, there would not be any central peaks within the crater floors.**

The outer perimeter of the crater was also peppered with smaller masses of material as it shot out from the center of the crater to the pressure ring at the edge. At which point the material shot into the sky a short distance to pepper down, and form the small craters in the rim (Fig. 104).

Around the rims of Alphonsus are many small craters. It has been stated that the craters were too many and too small to have been formed by impacts of meteors in these narrow, mountainous ridges. Yet, when one realizes that Alphonsus was formed by the impact of *mud* with the resulting secondary splash creating the small ridge craters, the process becomes quite clear.

Fig. 104. Note the peaks in the center of the craters created by the compact pressure zones during impact.

All laboratory experiments by others attempting to simulate the processes of crater formation have failed. The reason for failure is simply because a number of basic assumptions regarding the Moon and its origin and structure are wrong. There are a number of erroneous conclusions resulting from conjecture and deductive reasoning based on these assumptions.

For – the Moon is not billions of years old as is assumed. The craters of the Moon are not made up of volcanic lava or meteoric debris from outer space. It was never in a totally molten condition. It has no internal heats continuously generating, so it cannot have "liquid magma" in its interior to spew out over its surface. The Moon was not formed by the condensation of dust and gas at the time of the creation of the solar system. Neither is it a coalescence of space fragments that was later captured by the Earth eons ago. The Moon was never bombarded by great metallic or stony meteorites spaced millions of years apart. Nor are there volcanoes on the Moon, for the heat source of volcanoes is absent in its structure.

Ordinary strata of sedimentary origin containing organic materials is the only answer to the Moon's surface composition and is also the reason why lunar surface material reflects light as it does (7% of sunlight). And it seems that common mud containing organic debris has never been tested in the laboratory. One need not create exotic materials such as "fairy castles," foamstone, filimentary mineral matting, duststone, and the like, to solve the problem, for the moon is nothing more than a great chunk of the Earth's outer crust – and only plain ordinary soil will be found on the surface (except in the marias).

The Moon's surface was doubly covered in the vicinity of the craters with material of the late Cretaceous era – crustal deposits as well as plant and animal life. The Moon, composed of late Cretaceous crust, already covered with plant and animal life, was then drenched with additional deposits pulled from the remaining surface of the Earth.

Ultraviolet and infrared photographs of the central peak of the bright crater Aristarchus reveals minerals with possible sulphur content. Sulphur was very common in plants and other life forms in pre-catastrophe times. (Note the sulphur content of coal and petroleum on Earth.)

Mare Orientale is the largest crater on the Moon. This crater was created by the largest mass of very wet muds (about the consistency of a thick soup) which impacted the Moon at a very low speed. Large amounts of water flowed away from this crater in all directions as the muds settled. Great numbers of marine and plant fossils should be found in and around this crater.

NASA

MARE ORIENTALE

Fig. 105. The first crater to be formed on the Moon. The mass that created this impressive formation was almost totally water arriving from Earth at relatively slow speed, which, on impact, created waves. These waves left three rings around the center of the crater. Note the numerous flow-lines left by the water as it drained away from the outer rings.

Fig. 106. **Mount Wilson and Palomar Observatories**

The albedo of the lunar surface varies drastically in places. Some of the dark areas have an albedo of only 5% which points to an extrusion of igneous rock of shallow origin, whereas, areas resembling granite follow, since the mountains are extrusions containing granite and other light colored rocks. Many patches, noticeably around the craters such as Aristarchus, Tycho and so on, are quite brilliant with an albedo up to 35% or 40%, indicating, of course, salts accompanying sea water that arrived and formed the crater. Some of the last material to strike the visible side of the Moon were a few writhing masses of liquid rock which came from the deepest exposed layers – the quartz-granite (13th and 14th) layers of the Earth. This material was also the last to leave the Earth. Made up of very dense, molten rock, these liquid-like masses did not have time to cool appreciably before encountering the Moon. Upon impact, great splashes of liquid quartz-granite rayed out over the surface. Grimaldi, Eddington and Cruger are examples. The eastern limb does not reveal much evidence of molten rock impact because it was facing away from the point of impact when these masses were arriving. The fact that these impact areas overlap other features indicates and proves the late arrival of this material.

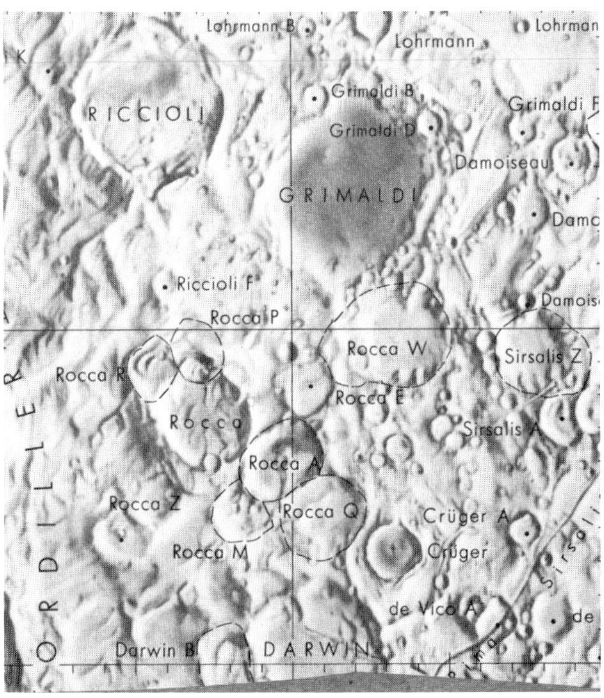

Fig. 107. **U.S. Government Printing Office**

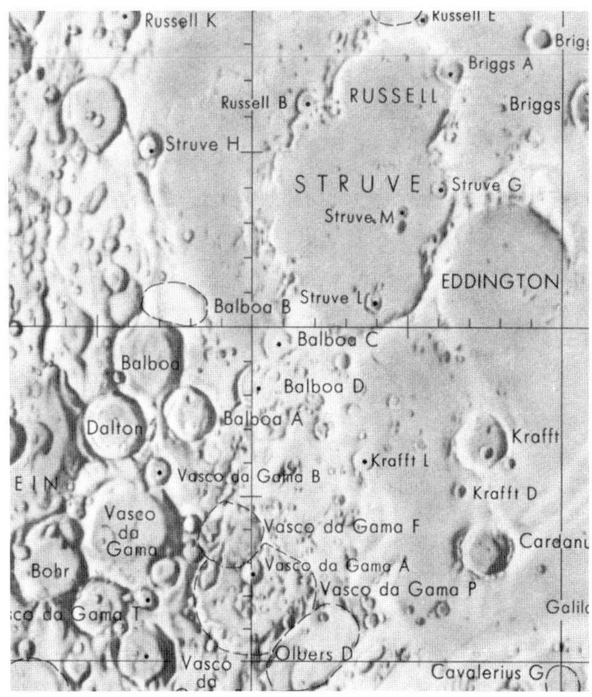

Fig. 108. **U.S. Government Printing Office**

It should also be noted that the largest group of small craters are in the central area of the Moon's surface. This indicates that they were also late arrivals attracted by the Moon to its center while it was in the final phase of moving into its position facing the Earth.

Without the influence of Earth-like atmospheric conditions on the Moon, there have been no surface changes of any magnitude since the Moon was created. Surface changes from meteorites and solar material have been slight and nearly inconsequential. A measurement of undisturbed dust in the maria basin (away from any craters) will give the rate of material arriving from the Sun annually since the catastrophe as well as the kind (composition) of material.

The only serious meteorite impacts occurred immediately after the Moon was pulled from the Earth and while its bowl-like shape was still partially facing the newly formed Asteroid belt. And then only after the attraction of the passing star had waned considerably. It was then that thousands of varying sized chunks and pieces of material from the other planets peppered onto the Moon's surface. Fewer of these meteoric impacts occurred on the far side of the Moon since the "clean-up" of all the fragments had been accomplished before the Moon had become spherical – before there was a "far side."

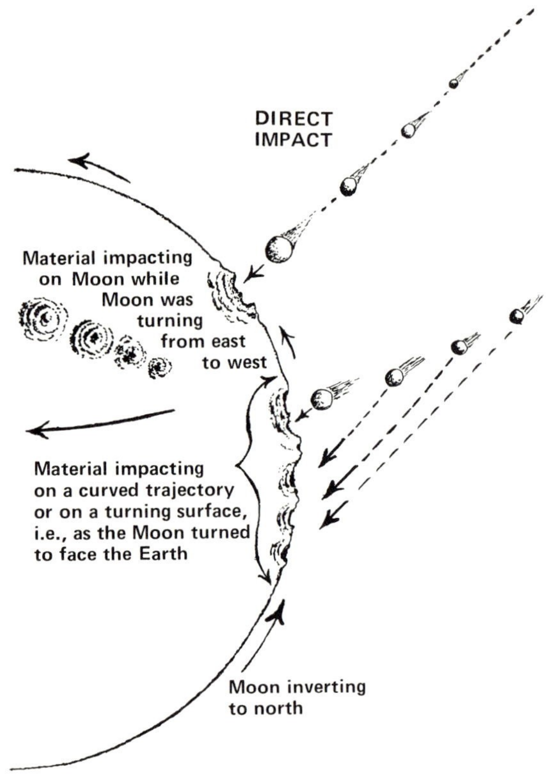

Fig. 109. Why some craters are located in "chains."

If the craters had been formed by meteorites from the Asteroid belt a spectrograph would reveal their metal content in the area of the crater, but no metal has ever been detected in the craters on the Moon. Also, meteorites do not, nor would not, land on the Moon's surface facing the Earth simply because the Earth, being a larger body, would draw all such meteorites to it instead. A few meteorites could land on the Moon when it is between the Sun and the Earth – when the Moon is in the path of a meteorite attracted by the larger body; when the Moon is between the Earth and the Asteroid belt, or when it is "full," meteorites may strike the far side, since they originate from the Asteroid belt or from comets.

One can be sure the material striking the Moon's surface was soft and wet rather than hard and stony for a number of reasons. Solid hard masses striking the Moon would have created very deep craters with small diameters and high rims without central peaks, whereas, all the craters have low rounded and sometimes collapsed rims.

The number of craters that will be found on the far side of the Moon will indicate the number of water masses that left the Earth prior to the crustal removal. One group, mainly water, preceded the Moon's removal, and the others, containing different soft surface materials following it.

The material that left the Earth first, was the last to land on the Moon. This material had traveled farther into space than any other material. Its return was slow because the pull of gravity by the Moon was weak, but the span of time was sufficient for the Moon to have reformed into a sphere before the majority of these late arrivals landed on the far side.

Upon close investigation, the Earth side of the Moon's surface will be found to be made up of more igneous material than the far side, since during involution it was the closing side, so less organic – or surface crust –will exist on the visible side. The surface crust that does exist will be extremely folded and broken, and intermixed with liquid rock. The new surface will be peppered with craters. There will be many broad and deep valleys and massive rolling hills. In a few areas will be found "sinks" caused by the final cooling of the liquid rock. Some of the recent pictures taken of the far side of the Moon show large depressions 125-300 miles across. These depressions confirm the involution process.

Fewer larger chains and craters and fewer "seas" exist on the reverse side of the Moon. This is true simply because involution occurred on the side facing the Earth. This side is mostly igneous material. And the larger crater-forming masses landed on the Earth side, whereas, smaller dispersed material landed on the far side.

Fig. 110.

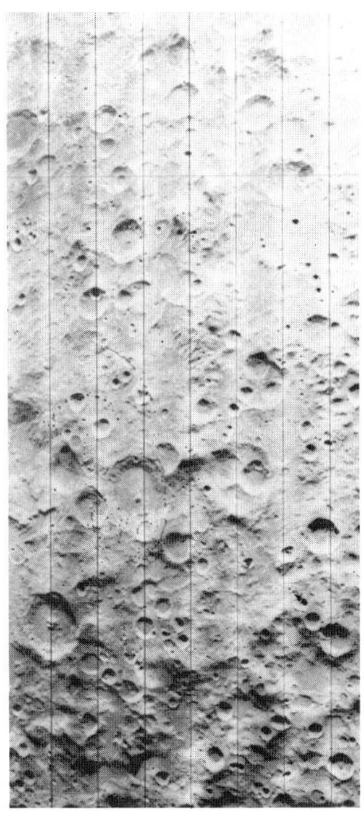

Fig. 111.

NASA

Fig. 112.

Fig. 113. Moon crater experiment conducted by Noel Dennis and Robert Landman based on theories of Lewis A. Manson.

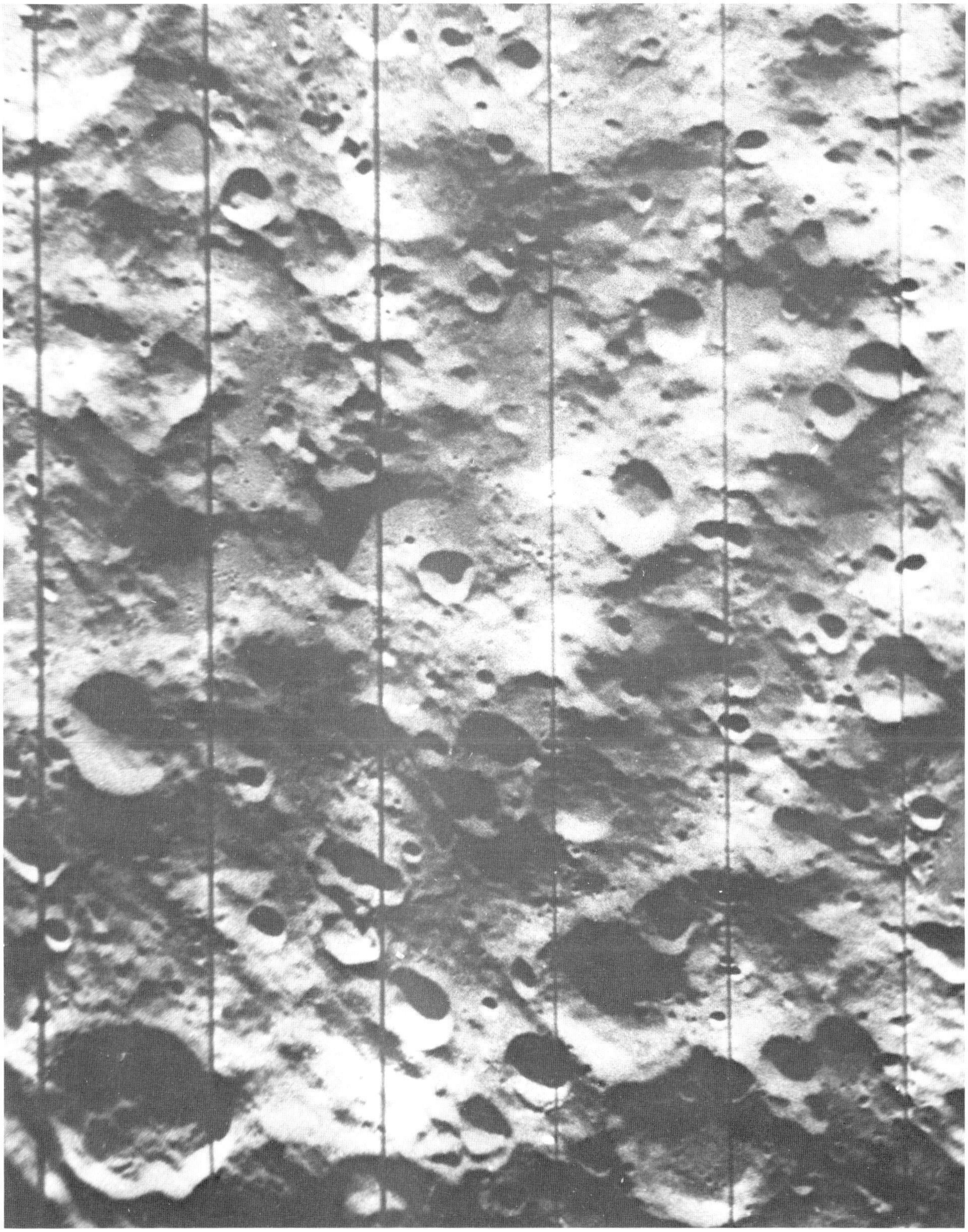

NASA

Fig. 114. Another Moon crater experiment? No — a photograph of the farside of the Moon by Lunar Orbiter I.

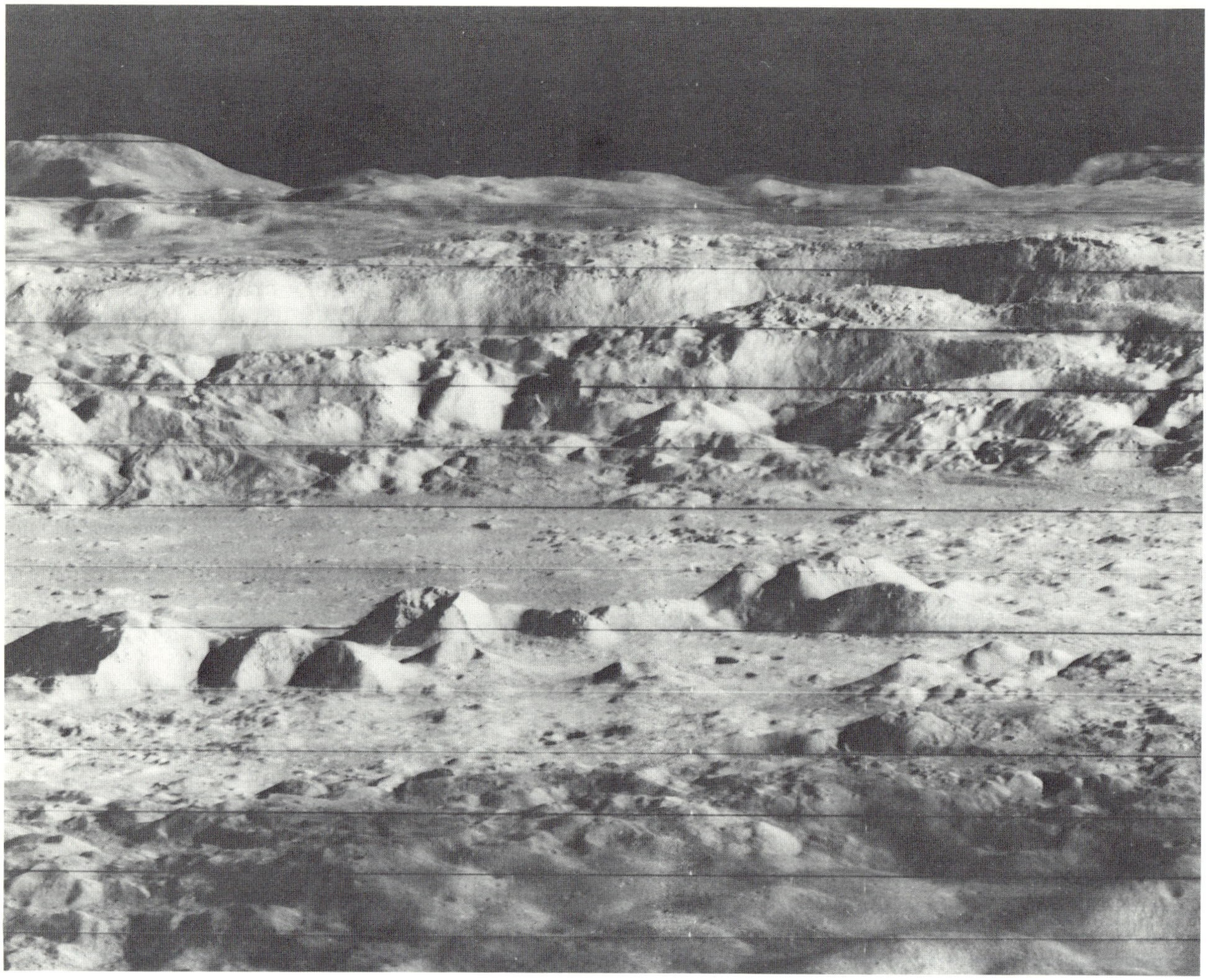

Fig. 115.

THE CRATER COPERNICUS

NASA

Closeup photograph of Copernicus taken November 25, 1966 by Lunar Orbiter II. Looking north from the crater's southern rim, the mountains that rise from the floor of the crater can be seen. These mountains rise 1,000 feet from the flat crater floor. The mountain on the horizon is part of the Carpathian Mountains. Cliffs on the rim of the crater are 1,000 feet high. From the horizon to the base of the photograph is about 150 miles. Lunar Orbiter was a little over 28 miles above the surface of the Moon when this picture was taken.

Lunar Orbiter was a National Aeronautics and Space Administration Project.

The Mysterious Rays and Lunar Mountains

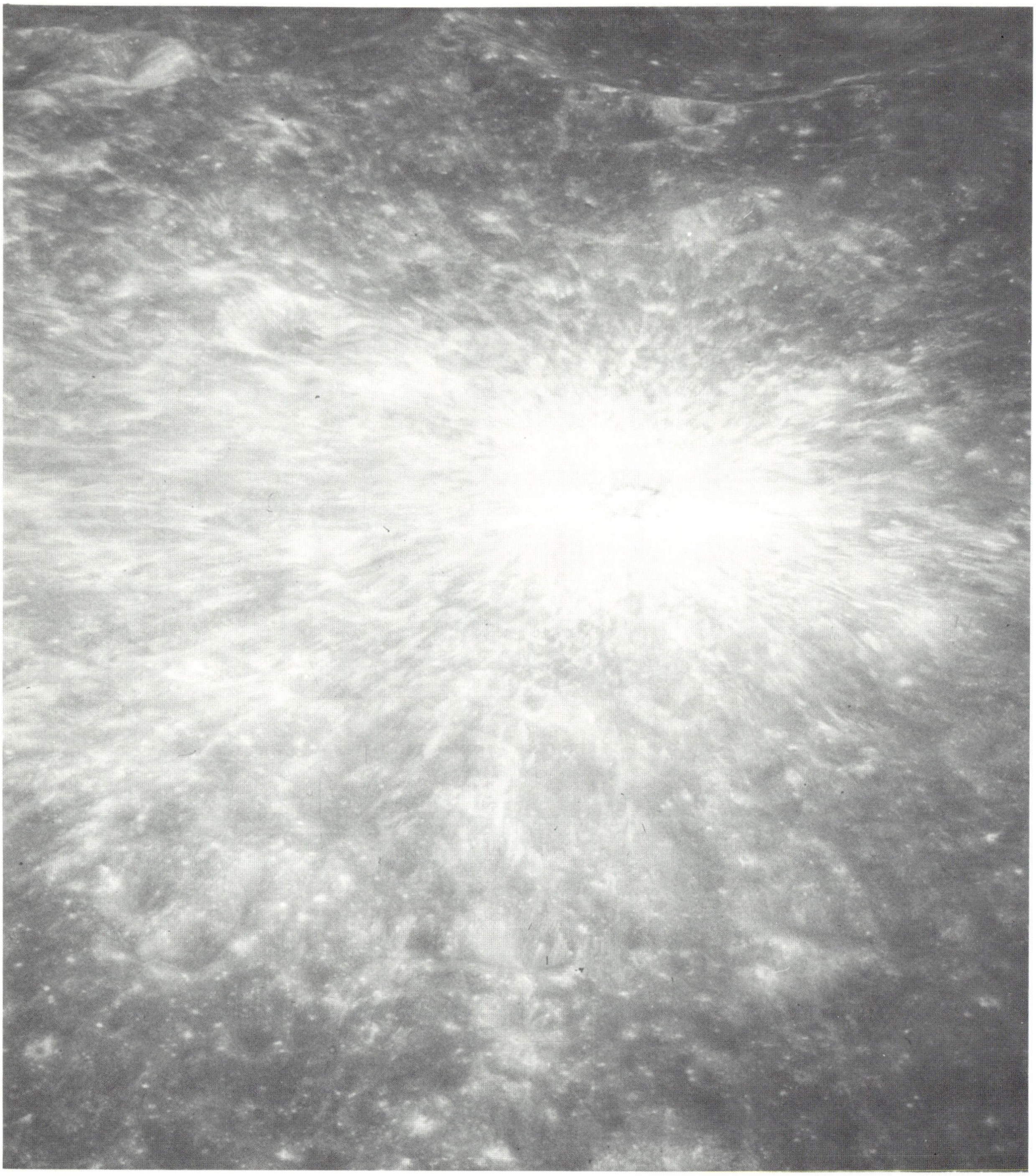

NASA

Fig. 116. A rayed crater formed by plumes of salt from the initial impact that rayed out for many miles in the vacuum.

Chapter 6

THE MYSTERIOUS RAYS AND LUNAR MOUNTAINS

The mysterious and famous "rays" emanating from such craters as Tycho, Copernicus, Kepler, Aristarchus and Obler were formed by a few straggling masses of almost pure salt. These great bodies of salt came from large quantities of sea water that had been pulled away from the Earth at the onset of the encounter with the passing star. This water was the first material to leave the Earth and the last to impact the Moon. It traveled far out into space in an attempt to follow the star, but soon lost momentum and was forced to return toward the Earth/Moon gravity center. During this interval the water evaporated into the vacuum of space leaving behind only masses of moist sea salt.

When these great globes of salt slammed into the surface of the now spherical Moon, the material rayed out in the vacuum from the point of impact for hundreds of miles in all directions (Fig. 116). The many smaller-rayed craters were created in the same manner by smaller quantities of salt. Since this salt was the last material to strike the Moon, it showered down and streaked over all other features, splashing them with a bright coating but not obliterating them. Because this material is salt, it easily explains the rays' high reflective quality when struck by oblique illumination during first and last quarter, yet when in full Moon are inconspicuous. Moisture in the lunar soil causes the salt to continually rise above arriving dust, maintaining the evidence of the original salt splash.

The rays belonging to Tycho begin a number of miles beyond the rim of the crater; they do not extend into the crater. This is simply the result of oversplash when the crater was formed. The newly formed rim prevented the splash of salt from landing on or into it.

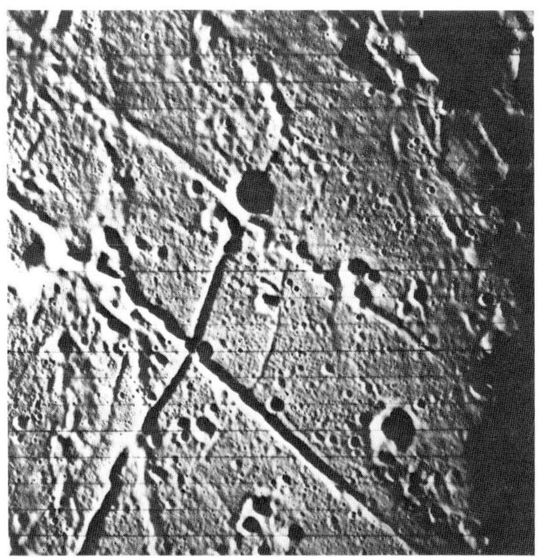

Fig. 117. Grid pattern on floor of crater Hevelius in Oceanus Procellarum. NASA

The plumes found at the ends of the rays are due to fine salt dust. In and around these rayed craters will be found quantities of dehydrated animal and/or sea life that had been carried along with the water.

A coarse grid, or criss-cross pattern has been seen over the face of the Moon; also recent lunar probe pictures have revealed numerous smaller grid patterns in many places. These are nothing more than multiple faulting of the crust during crustal distention and involution. The bulging floors seen in some clefts were caused by liquid rock that rose up through these fractures soon after they were formed (Fig. 117).

The steep mountains at the libration zones were formed by the involution activity as it created the spherical shape of the Moon. The Leibnitz and the Dorfel Mountains were created during the second series of major adjustments. Great spurts of liquid granite shot high into the crust during crustal distention and cooled in places, leaving jagged mountains standing some 30,000 feet high after subcrustal drainage had ceased and the crust had finished settling down around the new mountains.

Fig. 118. A flat plain becomes a great mountain as the sedimentary crust (3) is suddenly uplifted by enormous pressures which drive up liquid granite from the depths (2) followed by basalt, feldspar and deeper fluid rock (1).*

The line of "lunabase" areas found at the libration zone and which appear as though a portion of the surface had caved-in, were caused by subdrainage of liquid rock as involution was in progress.

One of the most important factors confirming the origin of the Moon and the method of its physical formation is the almost total absence of surface staining by metals. Few red, brown or purple hues exist on the surface; whereas, on the Earth these stains occur in the vicinity of every major fracture of the crust, proving the difference in the basic structure of the Earth and the Moon. The Earth is a many layered sphere of liquid and plastic rock and metal with a solid radioactive core, while the Moon is made almost entirely of quartz, granite and sedimentary rock, and contains a core of mainly granite formed from the 14th layer of the Earth.

There are no metal stains on the Moon except in the maria because the shallow crustal section was moved from the Earth so rapidly that it precluded the possibility of any metallic gas from the Earth's deeper layers from being injected into the departing crust. The Earth, on the other hand, was literally drenched with metallic gases of all kinds.

* Book of Mormon - 3 Nephi, Chapter 8, Verse 10: "And the Earth was carried up on the City of Moronihah that in the place of the City became a great mountain."

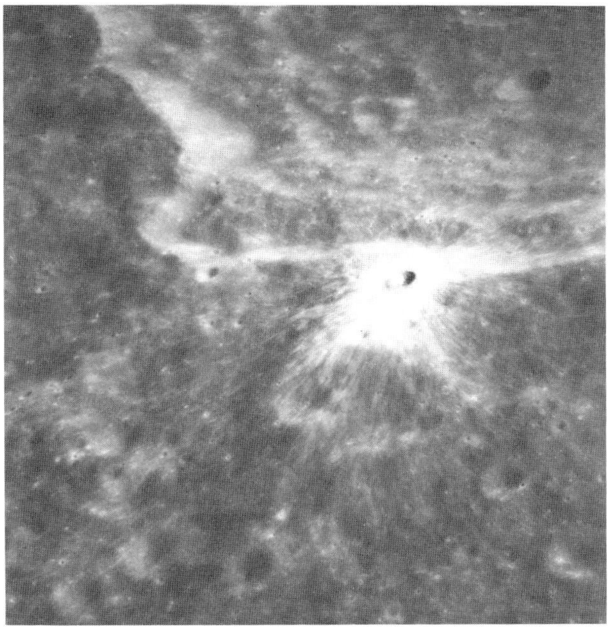
Fig. 119.

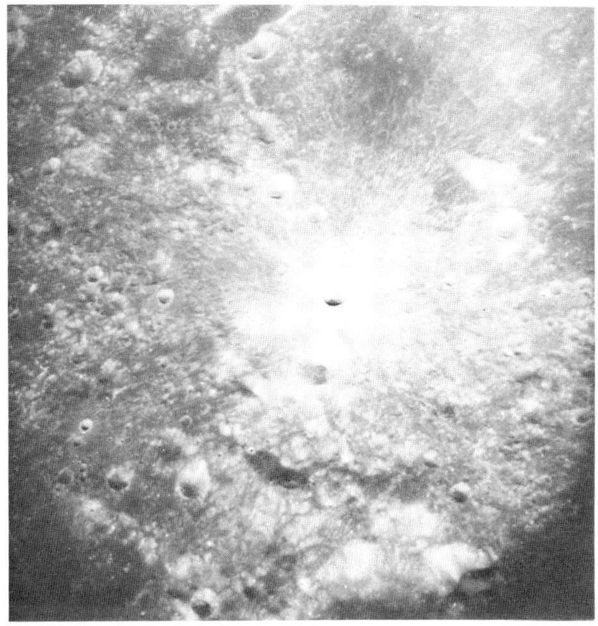
Fig. 120.

RAYED CRATERS

Fig. 121.

Fig. 122.

NASA Photographs

Fig. 123. NASA

MOUNTAINS ON THE MOON

Fig. 124. NASA

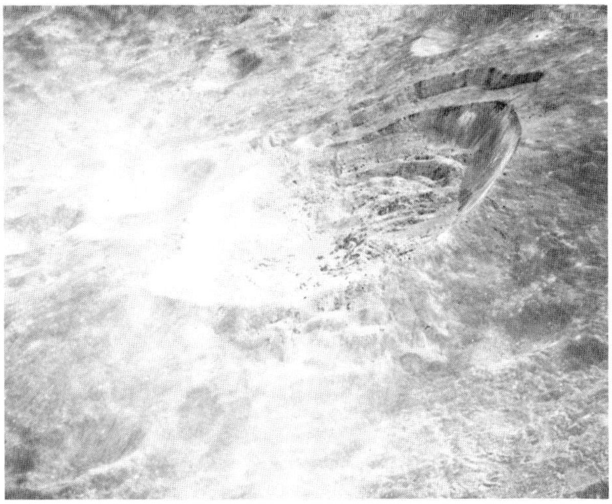

Fig. 125.

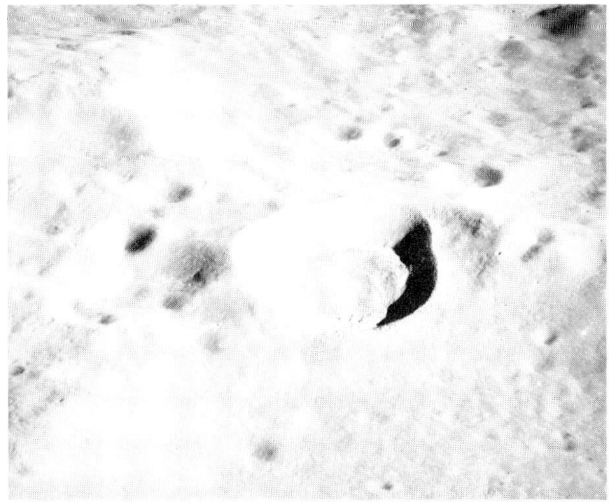

Fig. 126.

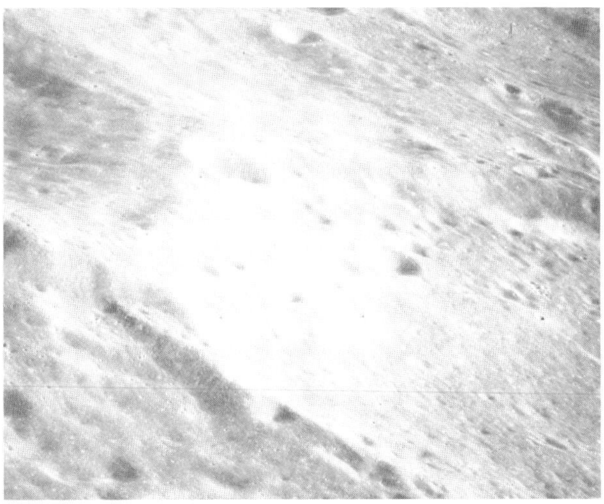

Fig. 127.

Fig. 128.

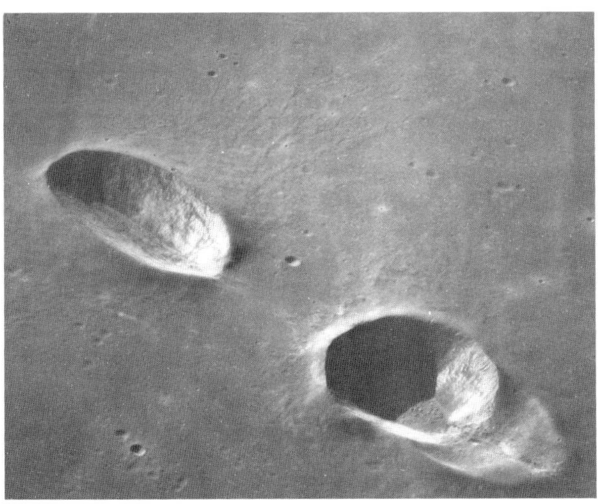

Fig. 129.

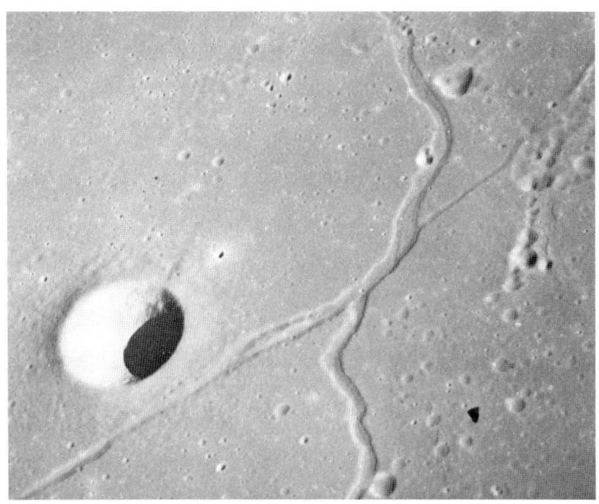

Fig. 130.

NASA Photographs

LUNAR CRATERS

Fig. 131.

Fig. 132.

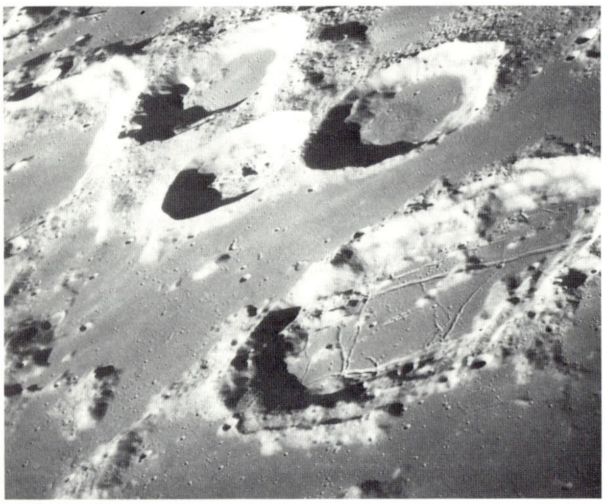

Fig. 133.

Fig. 134.

Fig. 135.

Fig. 136.

NASA Photographs

LUNAR CRATERS

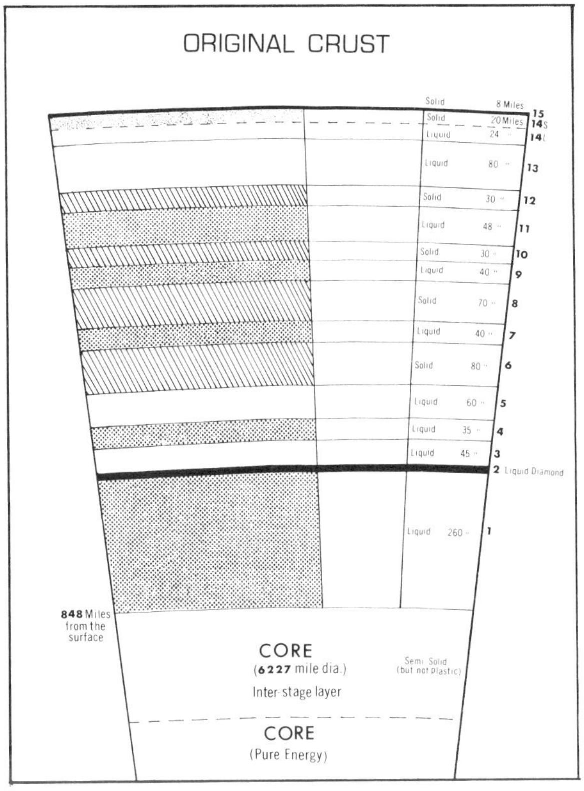

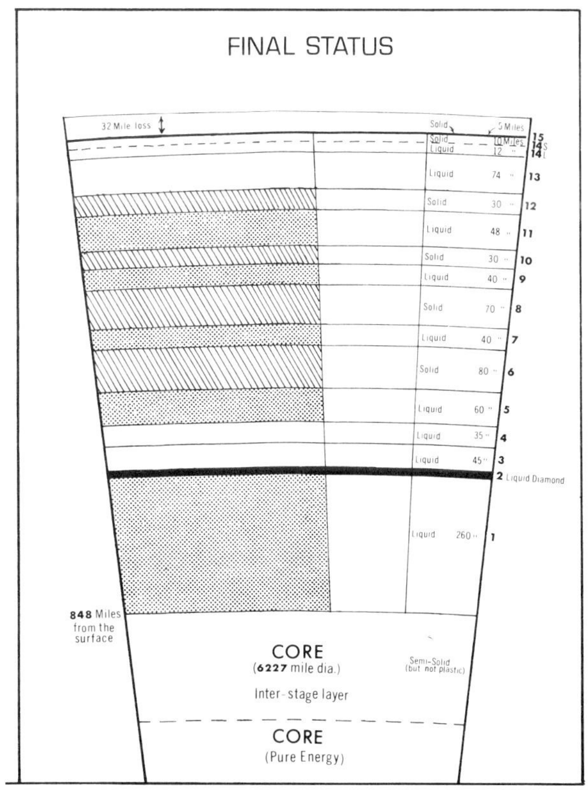

Fig. 137. These are cross-sections of the Earth from the core to the surface showing the original crust (left) and the present crust after the Earth readjusted following the catastrophe.

Fig. 138. NASA

The Mineral Deposits of the Moon

LUNAR MASCONS*

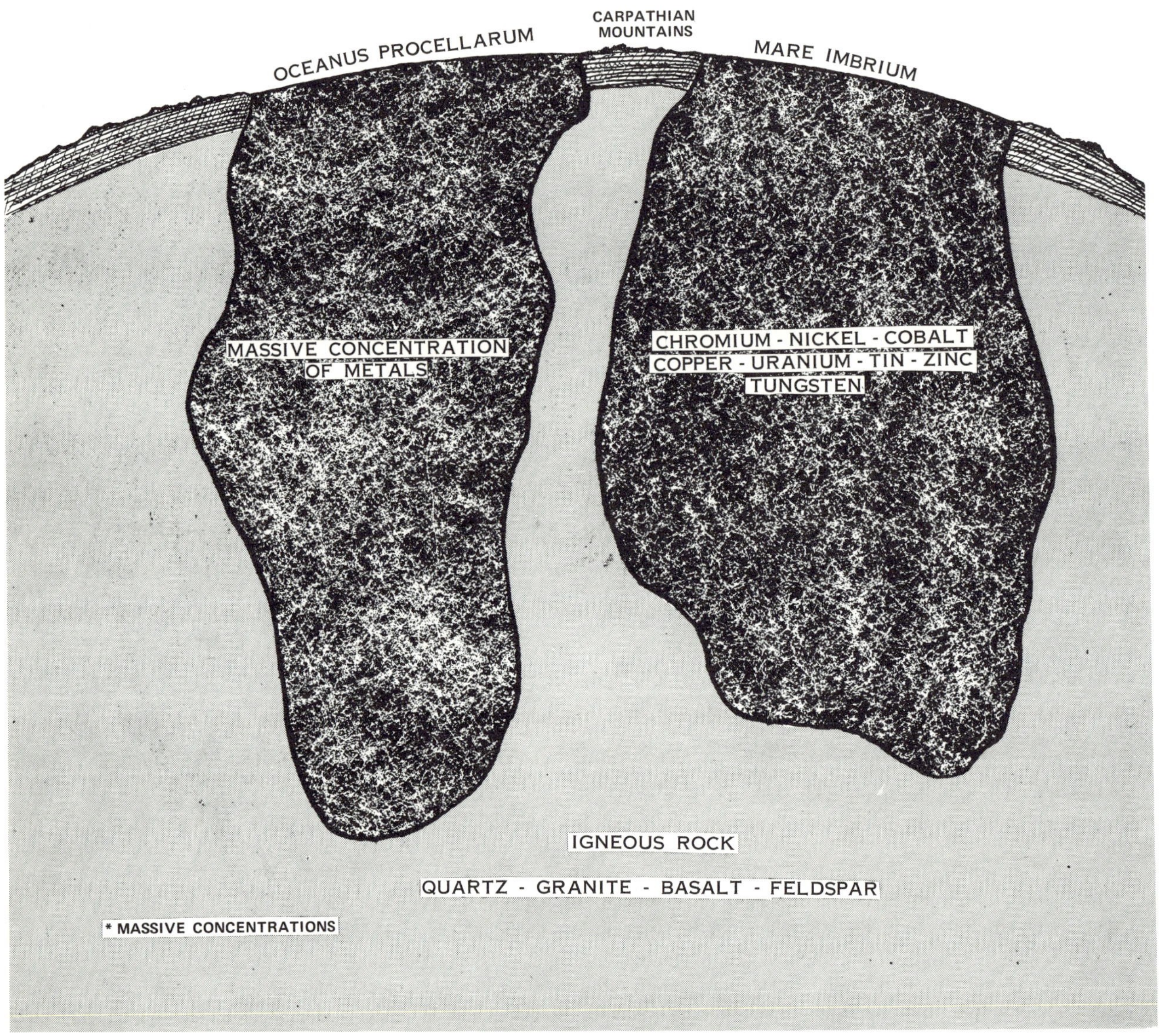

METALS ON THE MOON

Fig. 139. These metals were injected out of the Earth and into the Moon for a depth of about 600 miles at the moment of parturition. It is this great concentration of metals in four or five areas in the maria that prevented the rotation of the Moon. The Earth's gravity holds the heavier side locked in permanent position.

Chapter 7

THE MINERAL DEPOSITS OF THE MOON

As you look up at the Moon, little do you realize that you are seeing the largest mineral deposits on the surface of any body in the solar system. The gigantic deposits are in the marias, lying on the surface and extending down into the Moon for about 600 miles. The ore is extremely rich. The deposits would make a fantastic mining operation and supply the world's needs of chrome, copper, nickel, tungsten, cobalt, titanium, tin, uranium, zinc and very likely several other useful minerals (Fig. 139).

The Apollo 11 samples that were brought back to Earth contain beautiful glass beads, which were formed after splattering from molten rock as it was being injected into the Moon from the Earth's depths. These are mixed with the minerals on the Moon's surface. Some of the samples brought back consisted of titanium-bearing ilmenite, feldspar and a rainbow-colored rock called pyroxene. All of these minerals exist on the Earth..

The prevailing idea that the Moon coalesced out of dust and gases in its beginning is belied by the fact that these concentrations of metals in one small area could not have taken place if such was the mode of creation. Metals would be more evenly dispersed throughout the Moon. But there are no other mineral deposits on the Moon except in the maria. Also, sulphur, phosphorous, sodium and several other of the light elements will be found on the surface in large quantities *away* from the maria.

When the crust that became the Moon was removed from the Earth, it immediately began to close into a new sphere. It had an outer surface composed of soft, wet material eight miles thick, a semi-solid layer of quartz-granite below this twenty miles thick, a fluid layer of the same material twenty-four miles thick and twelve miles of highly fluid and very hot basalt, feldspar and orthoclase with pyroxenes. The crust was able to bend like taffy over the plastic and fluid rock below as it began forming a new sphere. The rock fluids within the new body immediately began forming a new core on departure from the Earth.

At the moment of departure, the gravitational attraction of the passing star had affected the inner metallic layers of the Earth to such a degree that the great vitreous layers that separated the metallic layers could no longer withstand the tremendous pressures being exerted upon them by the fluid metals and they fractured. These fractures are still evident in the Pacific Ocean basin off the coasts of the Western Hemisphere. The fractures are called Clarion Fracture, Marquesas Fracture, Murray Fracture, Demendo Cendo Fracture, Clipperton Fracture and the Cascadia Fracture.

These are the great fractures in the Earth's crust: the Third Vitreous Interbasement layer (12). Other lesser fractures in this arc are the Mendocino Fracture, Galapagos Fracture and the Eltanin. It was through these fractures that a gigantic mass of metals were extruded into the departing crust.

Although the flow was only brief (probably not more than five or ten minutes), it was enough to make the Moon very heavy on one side. Remember – the speed of the star was exceedingly fast. Thus, the pull of its gravity upon Earth was at its maximum at opposition for only a few minutes. Had the star passed just a few million miles closer to Earth, all the layers would have ripped apart and been destroyed. (If you think about it – all this activity during the catastrophe created a far better environment for man.)

On the floors of the Pacific Ocean, in the vicinity of all the fracture zones, lie great quantities of metals which flowed and splattered over many miles as the star's attraction waned. The greatest quantity of these will lie along the 30th latitude south. This is the center of the parturition point.

The Third Vitreous Interbasement layer fractured in hundreds of places in the hours just before opposition to the star. It was through these that all of the metals found today were injected into the surface crusts worldwide. (See pages 42 and 107).

As the newly formed Moon departed from Earth, the edges of the crust closed over the fluid rock and drowned in the edges of the maria. Very soon after – the arrival of the masses of mud, soft rock and other material occurred. It was these that created the craters and defaced the Moon's surface.

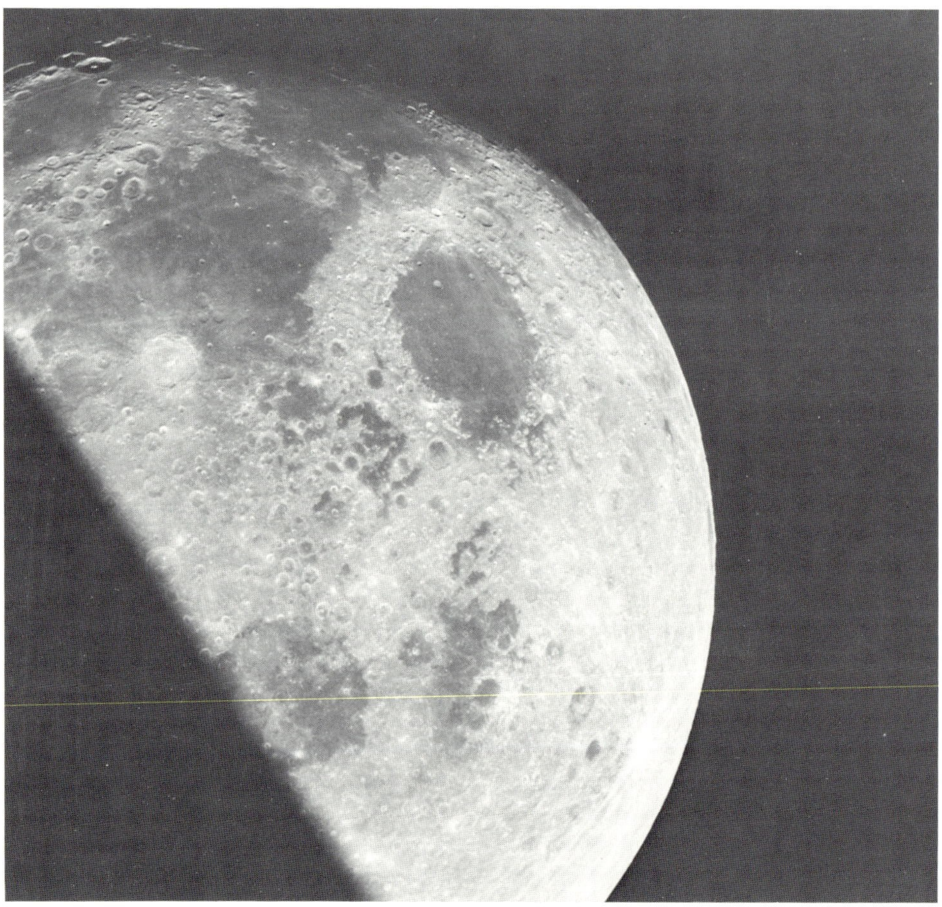

Fig. 140. NASA

The Moon Probes

SURVEYOR III (foreground) APOLLO 12 (background)

THE MOON PROBE

Fig. 141. NASA Photograph

Fig. 142. NASA

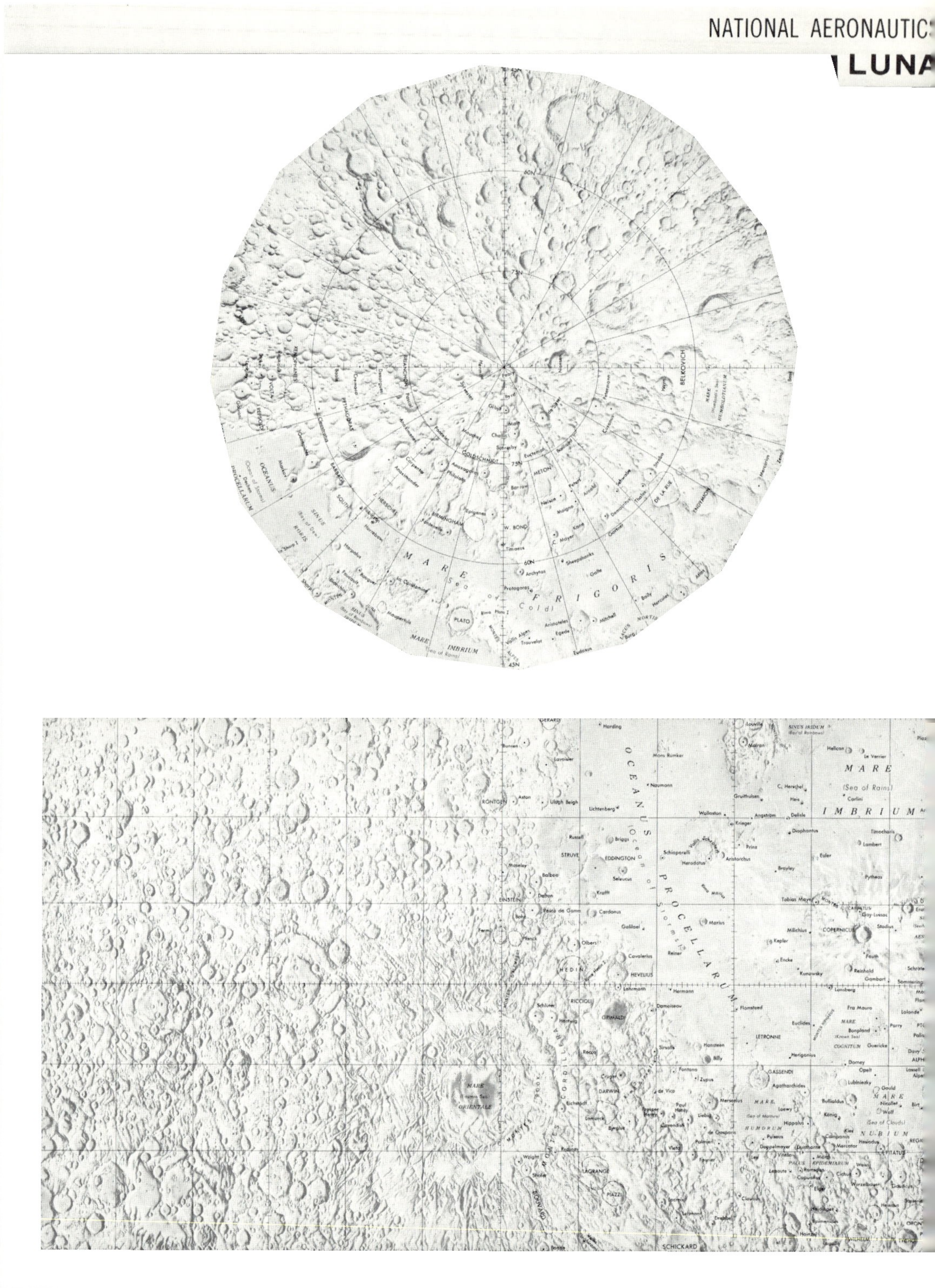

Fig. 143. LUNAR CHARTS NASA

ACE ADMINISTRATION

ART

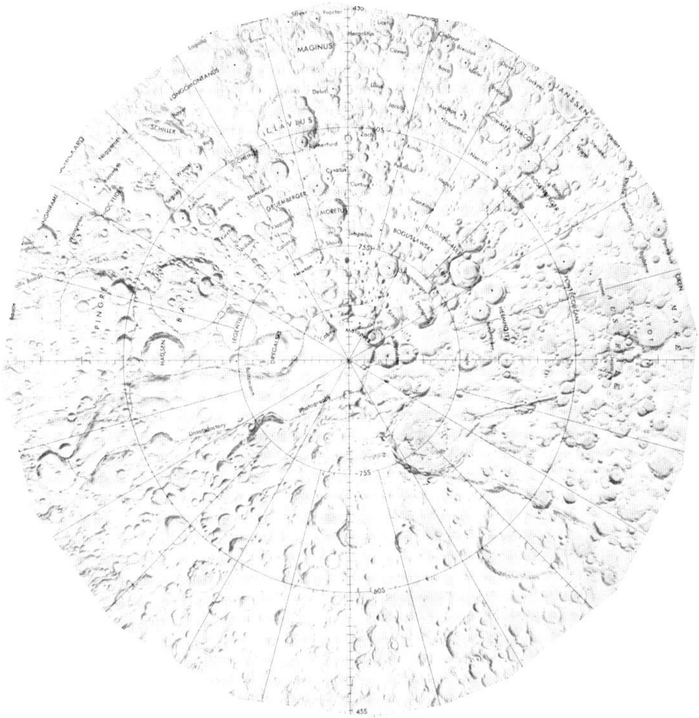

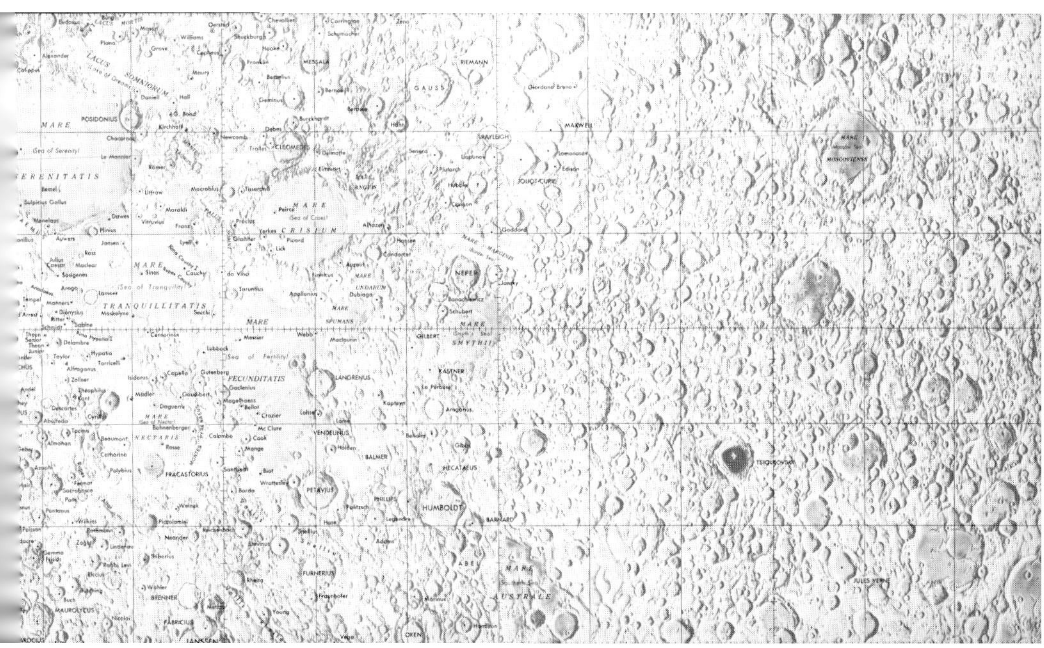

Fig. 144.

Fig. 145. See description on page 120 and Fig. 149. NASA

Chapter 8

THE MOON PROBES

The photograph made by Ranger IX at an altitude of 785 miles covers an area of 149 x 124 miles. Ptolemaeus, Alphonsus, and Albategnius are the large craters in the photographs. These were formed by Group I (Ptolemaeus) and Group II (Alphonsus and Albategnius) masses. Many smaller Group I blobs that impacted later are also well defined.

In the next photograph, Alphonsus is the only major crater in view. Within the crater, as well as on the rims and surrounding terrain, are numerous small impacts of shallow extent. These were created by secondary impacts of material returning from the vertical splatter from the rim of the crater (Fig. 147). The mass that formed Alphonsus caused hundreds of small "drops" to fly up from the pressure rim and, in dropping back to the surface, created the small craters all around the rim.

The masses that created these craters arrived during the moments of weak gravity field when the Moon was between the star, the Earth and the Sun. The area shown is 67 x 62 miles.

The next photograph was taken at an altitude of 8.3 miles and shows the floor of Alphonsus. The large depressions are not craters but "sinks" caused by subsurface adjustments. The small pockmarks are tiny craterlets produced a few minutes after the largest ones had been made. At right center within the sink are three aligned pockmarks. These were created by heavy solids of small size. The area this photograph covers is 1.6 x 1.4 miles.

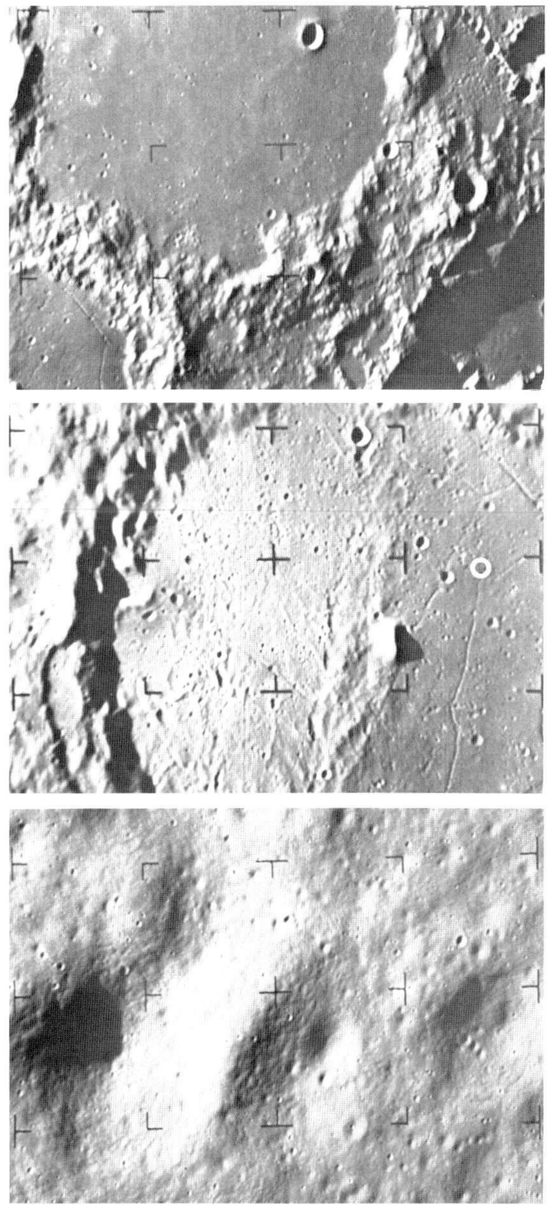

Figs. 146, 147, 148. NASA

At the time of this writing, I made an almost unbelievable discovery from NASA's Surveyor I, which sent back photographs of the Moon's surface, June 1966. Just a few feet from the cameras lay startling evidence confirming the facts detailed in this book.*

An object photographed by the cameras and which at first glance may seem to be nothing but a rock, is, as can be seen in the drawing, not a rock but the totally dehydrated head of a dinosaur. I have classified this dinosaur to be one of the "duckbilled"types known as a Kritosaurus.** These animals roamed the Earth in great numbers during the late Cretaceous Era. The head and other fragments lay within ten feet from the Surveyor.

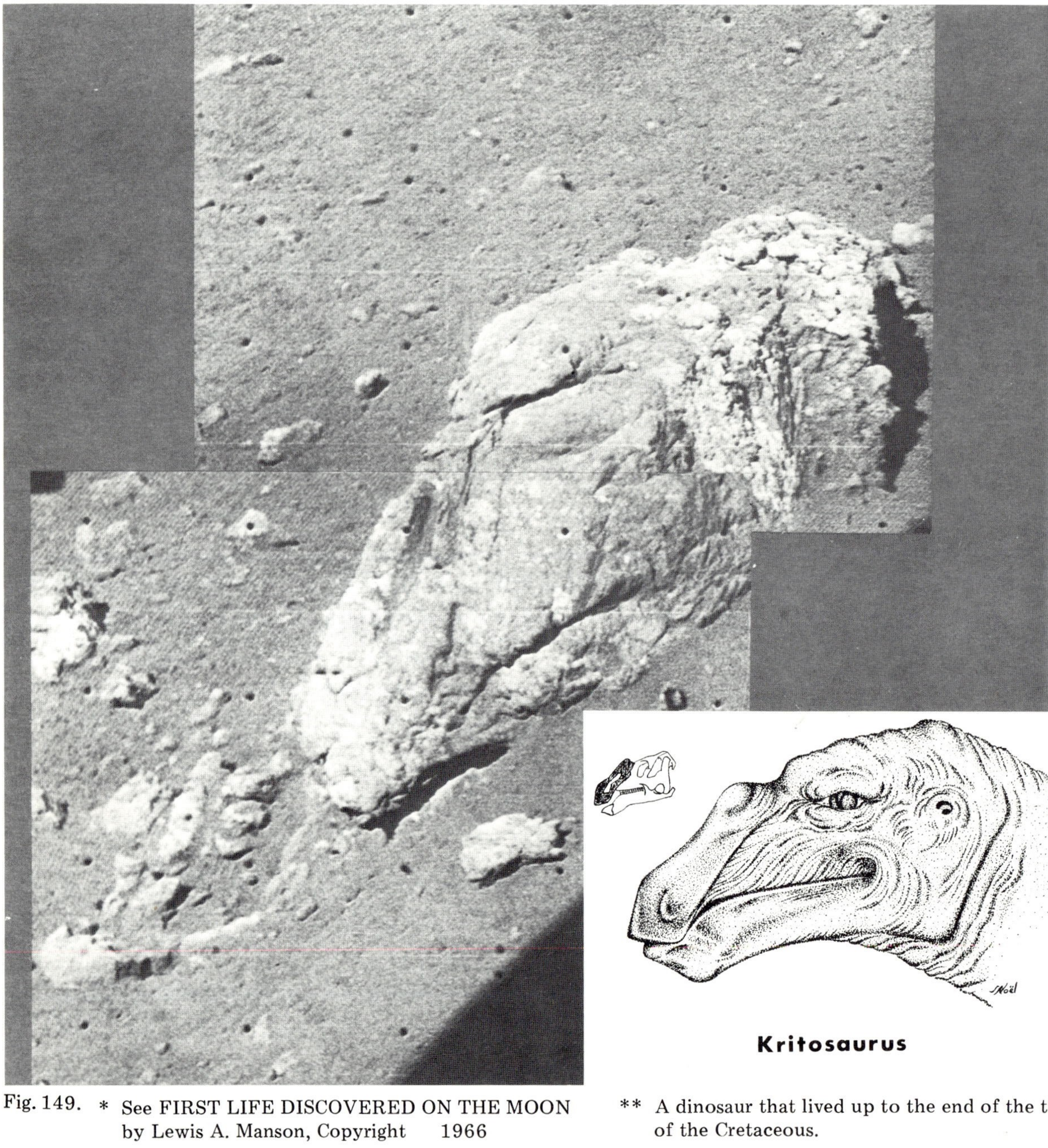

Fig. 149. * See FIRST LIFE DISCOVERED ON THE MOON by Lewis A. Manson, Copyright 1966

** A dinosaur that lived up to the end of the time of the Cretaceous.

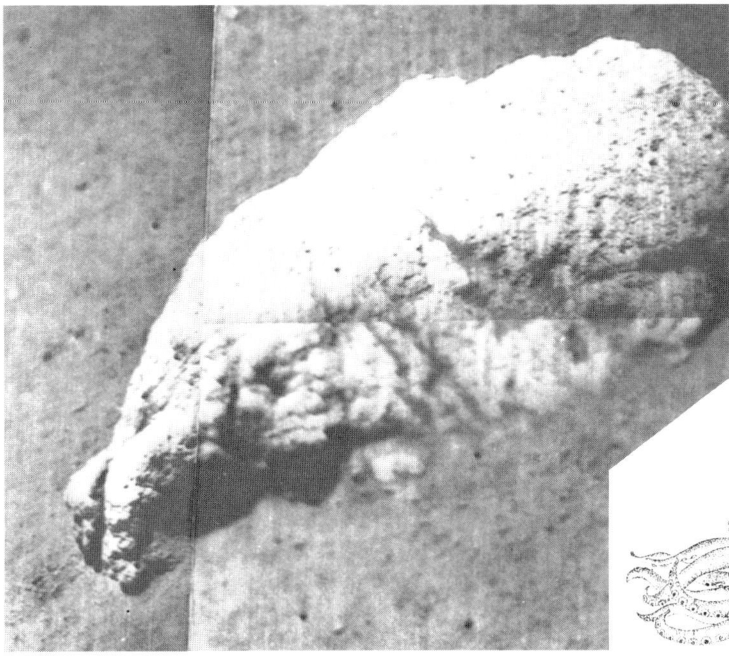

Fig. 150. **NASA**

Fig. 150 is a photograph of another odd shaped "rock" taken by Surveyor I, which I believe to be a cephalopod – fleshy parts only. It had been pulled forcibly out of its shell by the violence of the catastrophe, or the vacuum in outer space caused pressure within the shell to eject it.

The items mentioned in these photographs do not indicate in any way a rock structure since fracturing of sedimentary rock would not result in a curved or transverse fractured appearance. Igneous rock would not cool in this form for a number of reasons. Most of the remains of life found on the Moon were dashed to pieces by the impact of the incoming masses and rushing waters.

Without having true knowledge of the Moon's creation, it is easy to understand why objects lying near the Surveyor I were identified as "rocks." It was long believed that the Moon was a dead planet, made up of cold rock and lava which through the ages was continuously pelted by meteorites until its surface was a shattered, rock-strewn ruin. But if one will consider what the Moon really is – a section of the Earth's outer crust, the idea of dinosaurs and cephalopods lying strewn about on its surface is not inconceivable.

Evidence of former man on Earth is very rare as are the remains of smaller animals of all types, since all smaller life was readily devoured at death by the numerous larger animals – and the Earth teemed with them – that existed in the same environment. What was left was consumed by smaller animals, insects, bacteria, etc. for the Earth, seas and swamps were thick with life of all kinds.

Also, early man being in an extremely primitive state of consciousness did not yet practice the burial of his dead. He left such remains to the mercy of the elements and animals. It is, therefore, only by chance that bits and pieces of his skeletal remains are found in ancient deposits. However, much evidence of this early man will without doubt be found scattered over the Moon's surface. These remains will be in a totally mummified condition as was the Kritosaurus and cephalopod already noted.

Any form of life that existed on the Earth at the time of the catastrophe will be found on the Moon perfectly preserved, though much of it in a torn, bruised and fragmented condition. This includes all life – animals, insects, plants – down to the smallest forms. And unless solar radiation has effected them, any seeds or plants found on the Moon could be still viable if returned to a suitable environment for their growth.

There is enough gravity on the Moon for it to retain some form of atmosphere, with no likelihood whatever of these gases streaming off into space. Were its gravity small, the Moon could not have closed into its present shape. Granted, the atmosphere would be made up of the lighter gases: hydrogen, neon, oxygen and helium.

This atmosphere will hug the surface and seep to shadow areas and depressions. It will slither across the surface thin and wraith-like before the rays of the rising sun. This is partly the reason it has failed detection through telescopes and observatories from Earth. There are faint wafting breezes on the Moon's surface. The evidence of shifting dust particles were photographed by Surveyor I on the surface of its own polar panels (Fig. 151).

Fig. 151. NASA

The Earth's gravity holds the heavy side of the Moon toward the Earth. Though only one side of the Moon is always facing the Earth, the Moon does "seesaw" slightly when passing other bodies in space. This physical libration is caused by the Sun and the other planets in the solar system and where they are located in space as the Moon travels around the Earth. If, during its orbit, the Moon approaches Venus to the east of it, the face of the Moon which we see from Earth will tend to swing in the direction of Venus. Then as the Moon gets directly in line with the Earth and Venus, the Moon will face the Earth normally, for it is then directly in line with the two planets. But as the Moon passes on around the Earth and passes Venus – which now takes a position west of the Moon, the Moon will then swing west under the influence of Venus' gravity.

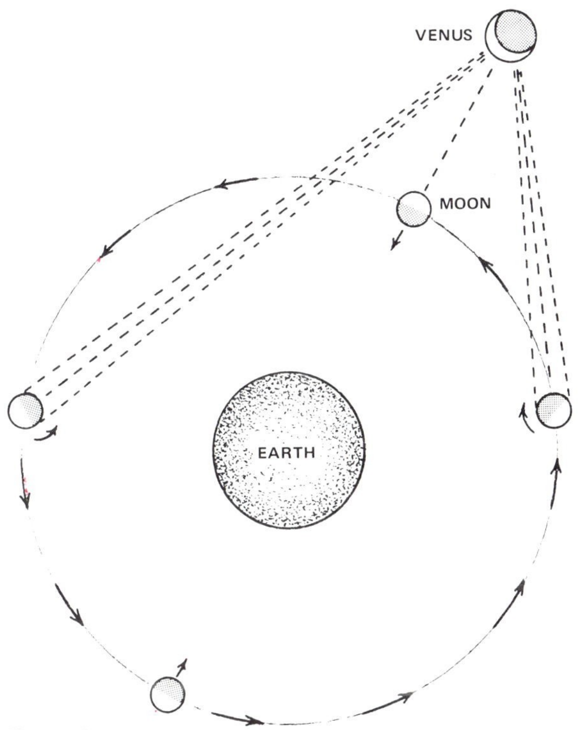

Fig. 152.

Surprisingly, and contrary to popular belief, the Moon does not create the tides of the Earth. Through instrumentation, we now know that it is the planet, Venus, that is the main cause of the tides, with Jupiter contributing the secondary major force. This has been established through new types of instrumentation I built and operated for several years.* The Moon has a small tidal effect upon the Earth but this tidal effect is actually that which changes the *time* of the tides as it orbits the Earth (Fig. 153).

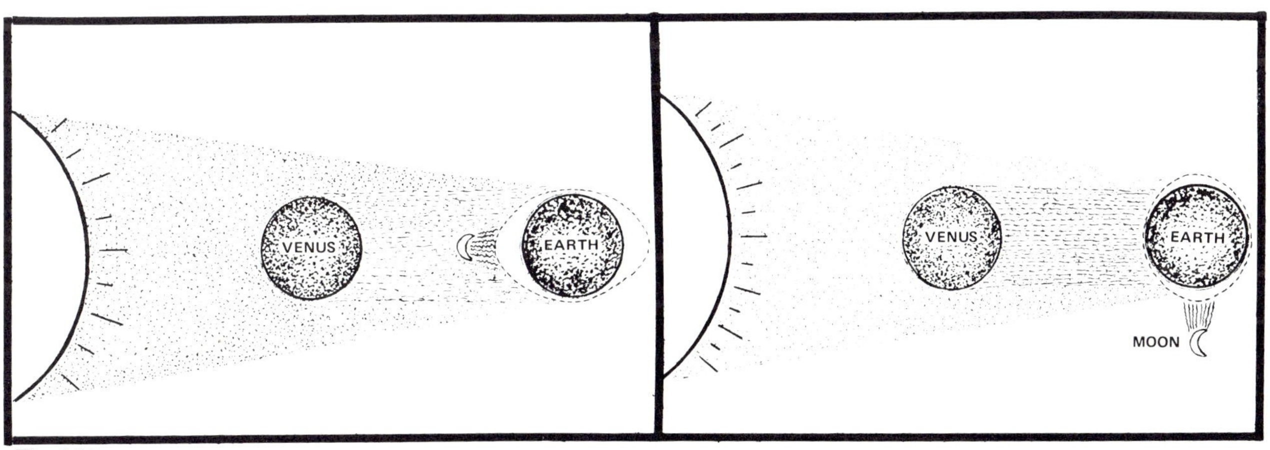

Fig. 153.

* See graphs on pages 128 thru 131.

The great gravitational attraction of Venus creates surges of pressure by drawing upon the liquid layers within the Earth's crust (the same as the great passing star did except on a minor scale), which then surge and meet at numerous points every day under the Earth's crust (Fig. 154). Upon meeting, brief impacts of pressure-surges create subterranean shocks of varying intensity. The planet Jupiter also affects the Earth's liquid layers.

Furthermore, tidal effects are not set up initially on the surface, which are thought cause the shifting of water of the oceans to shore and back; the water is set into tidal motion by the rise and fall of the Earth's crust *under* the water, activated by the liquid metals surging many miles beneath it which act as "springs"* under the surface of the crust (Fig. 155).

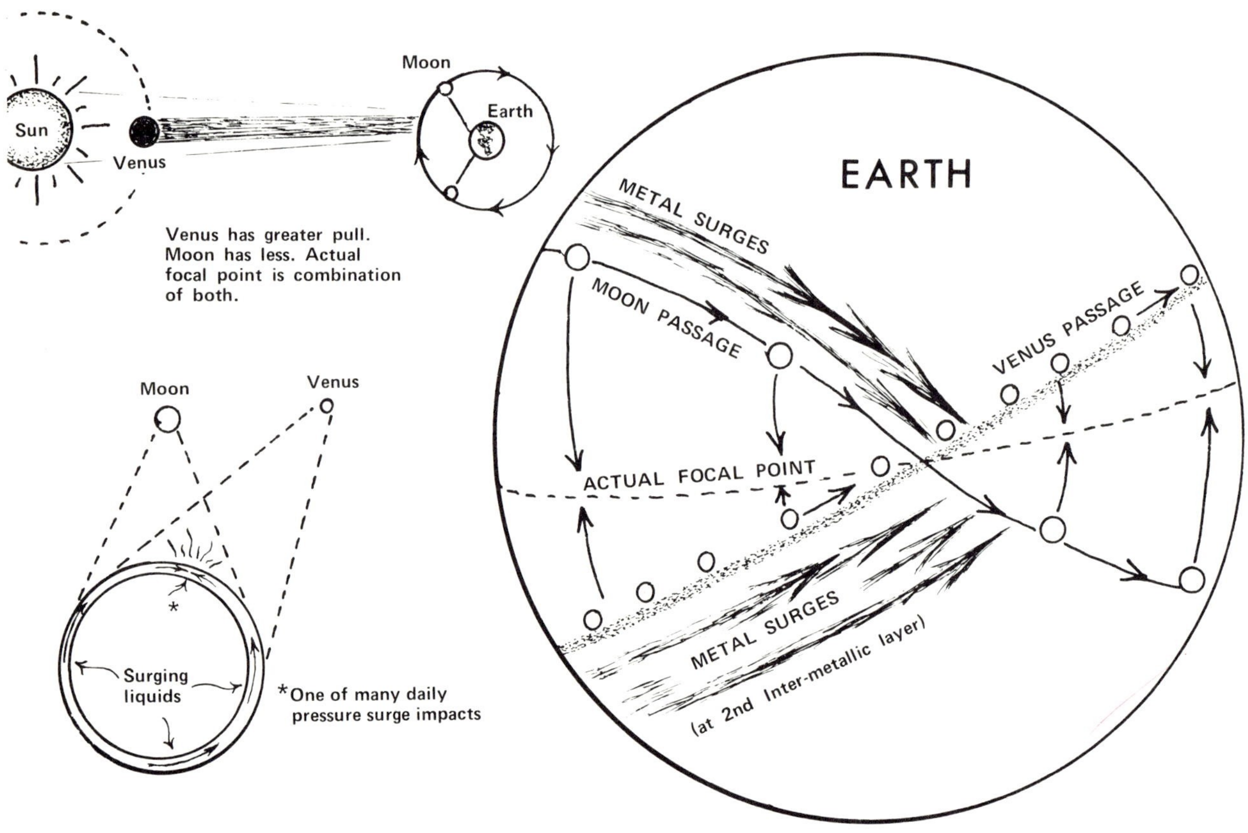

Fig. 154.

NOTE: There are no tidal effects on the Moon — bulges and other unevenness observed in its structure were not caused in this manner. The Moon is a solid, cohesive mass since it has no radioactive core to generate and maintain internal heats which would give rise to certain plasticity in the Moon's structure and maintain a fluid condition in some of its layers - if such existed.

* Job: 38:16

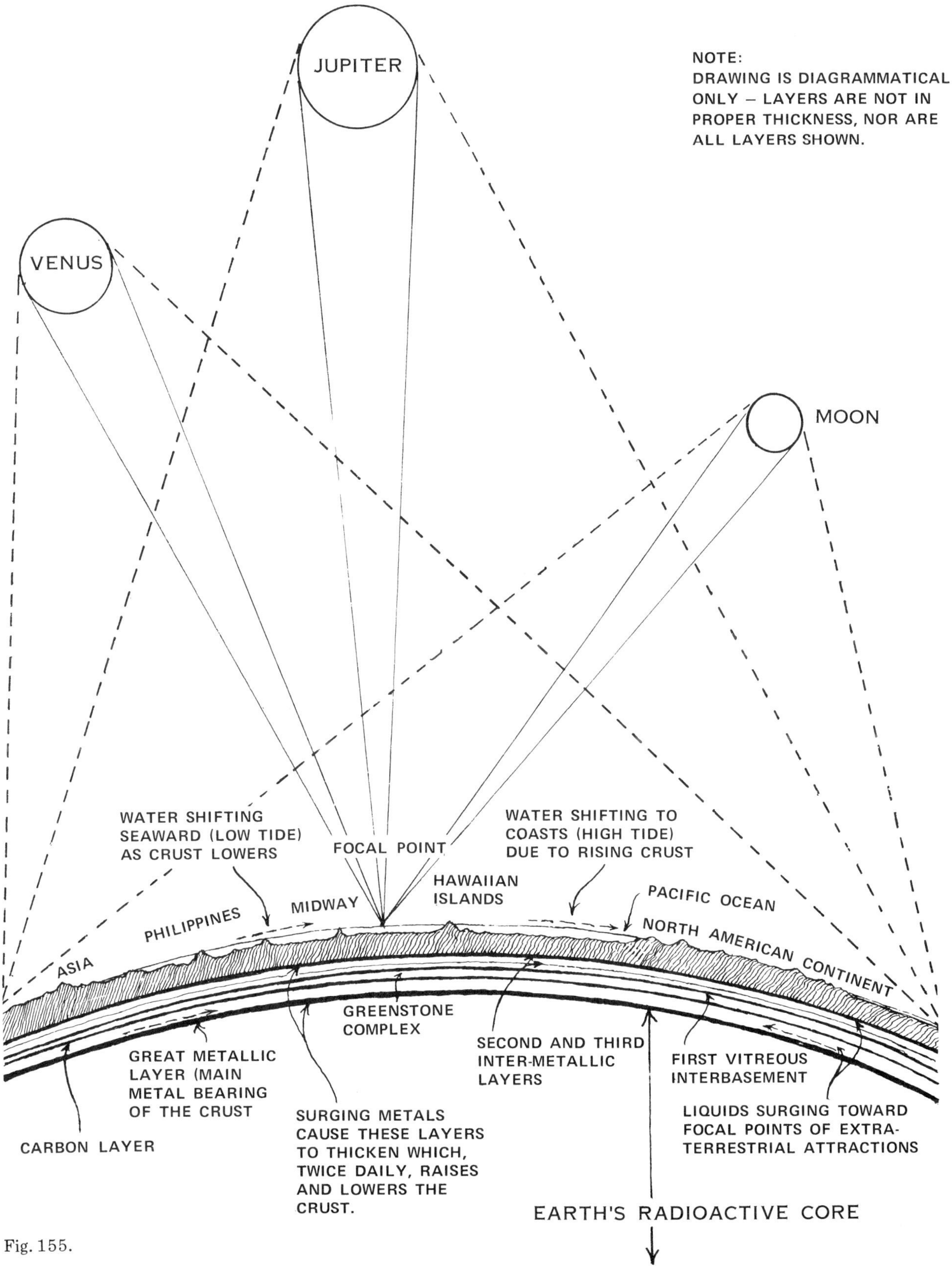
JUPITER
VENUS
MOON
NOTE:
DRAWING IS DIAGRAMMATICAL ONLY – LAYERS ARE NOT IN PROPER THICKNESS, NOR ARE ALL LAYERS SHOWN.
WATER SHIFTING SEAWARD (LOW TIDE) AS CRUST LOWERS
FOCAL POINT
WATER SHIFTING TO COASTS (HIGH TIDE) DUE TO RISING CRUST
HAWAIIAN ISLANDS
MIDWAY
PHILIPPINES
ASIA
PACIFIC OCEAN
NORTH AMERICAN CONTINENT
GREENSTONE COMPLEX
GREAT METALLIC LAYER (MAIN METAL BEARING OF THE CRUST
SECOND AND THIRD INTER-METALLIC LAYERS
FIRST VITREOUS INTERBASEMENT
CARBON LAYER
SURGING METALS CAUSE THESE LAYERS TO THICKEN WHICH, TWICE DAILY, RAISES AND LOWERS THE CRUST.
LIQUIDS SURGING TOWARD FOCAL POINTS OF EXTRA-TERRESTRIAL ATTRACTIONS
EARTH'S RADIOACTIVE CORE

Fig. 155.

If you will refer to the drawing of the structure of the Earth in cross-section on page 42, I will explain the layers in which earthquakes are generated. First of all, there are several kinds of earthquakes and generally each are generated in different layers. The 14th layer produces the minor temblors that are common worldwide daily. These are caused by differential pressures as the layers below continuously rise and fall.

From the 11th layer originates stronger quakes. Since the layer is only 48 miles thick, the surges do not generate enough pressure to create the major shocks. It is from this layer that most of the quakes that give an effect like a rumbling freight train running underground. This effect can also originate from the 7th and 1st layers when surges arriving from each side of the Earth fail to meet but pass each other with a side-swipe action.

The 11th layer produces earthquakes of the 3rd to 5th magnitude usually. The 7th layer is 110 miles thick and is extremely fluid. The temperature at this level is between 7,000 and 8,000 degrees Fahrenheit. It is from this level that the majority of major quakes originate (5th to 7th magnitude). It is also this level that responds continuously to extra-terrestrial gravitational forces and in which the major surges are set in motion.

Layer number 1, the main metallic bearings of the crust of the Earth, has a temperature ranging from the base of 6,000 degrees to 9,000 degrees at contact with layer number 2 – the carbon layer. The metals in the upper half of this layer are in an extremely fluid state and respond almost instantly to any force. This layer is the originator of the major and giant earthquakes. Seismic records of these quakes show depths of origin at 590 to 650 miles. The magnitude of these quakes ranges from 7 to 10 which is total surface destruction of man-made objects and distortion of the surface rocks.*

There are no moonquakes on the Moon since there is nothing within the Moon's interior to create seismic shocks because of the total lack of liquid layers that could respond to the gravitational attraction of Earth, Venus and Jupiter. Without these fluid layers to surge, there can be no tidal effects or moonquakes. The Moon is locked up solid into a completely frozen sphere. What internal heats there are have been slowly dissipating since the Moon's creation.

Fig. 156. **NASA**

*** This subject is covered in great detail in another work titled, "Why the Earth Quakes," in preparation.**

The graphs on the following pages are a continuous record of gravitational variations at a stationary point on the Earth's surface. These graphs were made from automatic recording equipment. The graph sections reproduced here are extracts from different times to show the widely varying cycles of deep space and the gravitational forces at play. These variations are due to the varying forces from the Moon, planets, Sun and the stars as the Earth revolves. You will note on the graphs that the planet, Venus, is always the greatest attractor. The sharp dips marked "S" on the graphs are rare but indicate unusual surges of the liquid layers deep in the layers of the Earth's crust.

The knowledge gained from this equipment has given us much new information on the true cause of earthquakes and the Earth's structure. But most important of all is the knowledge gained on gravitational force. Unfortunately, it is not possible to give greater detail about these graphs and what they reveal in this book.*

* Greater detail is given in WHY THE EARTH QUAKES in this series.

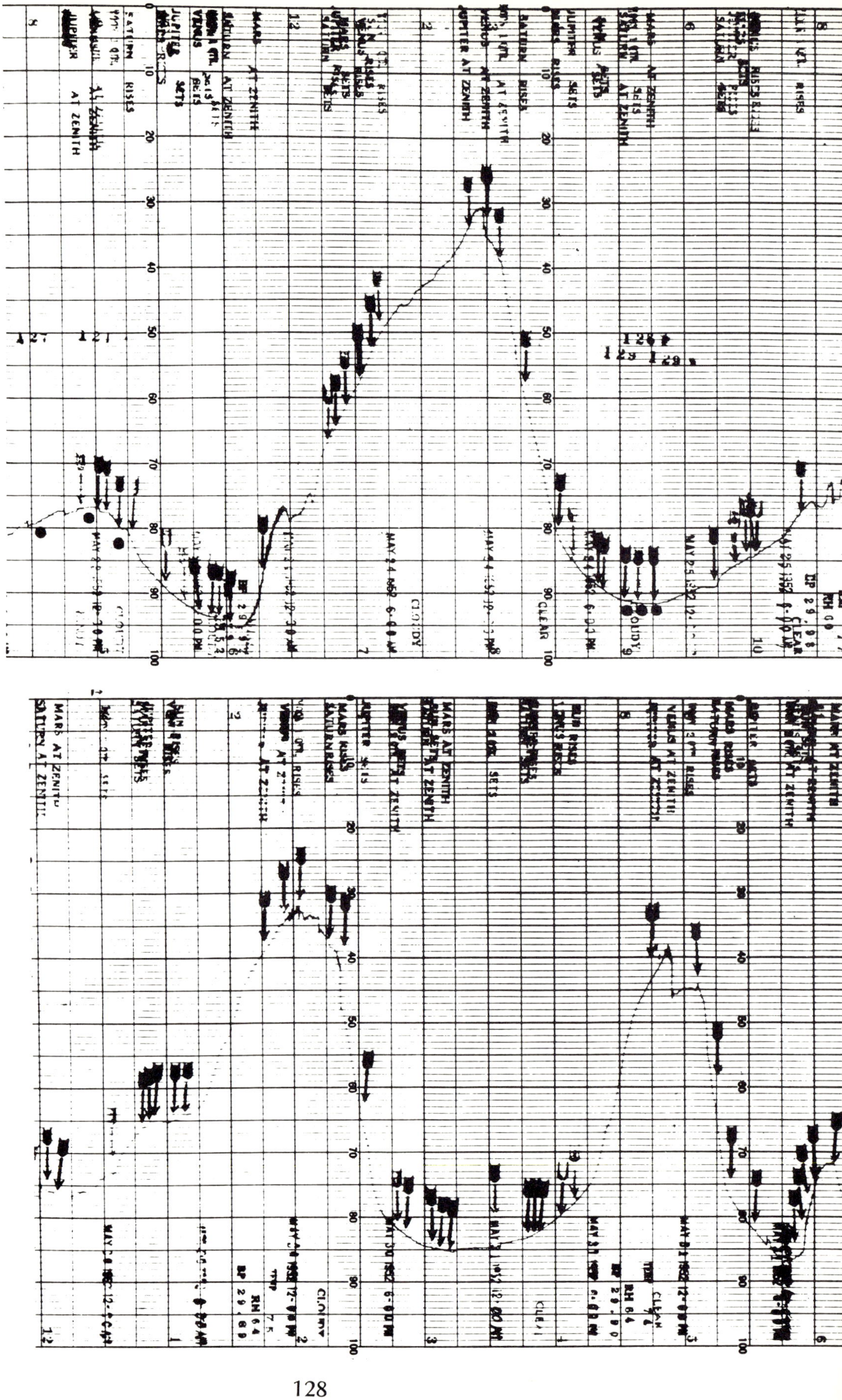

Fig. 157.

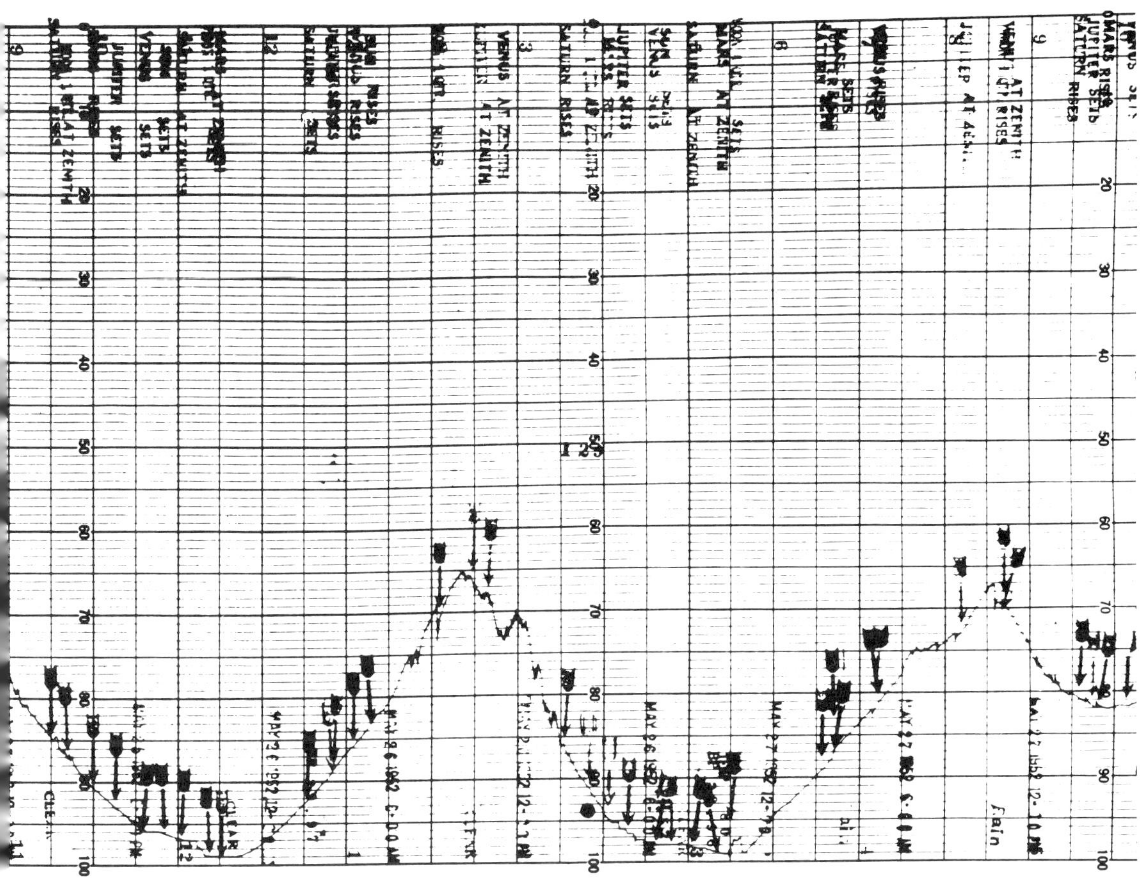

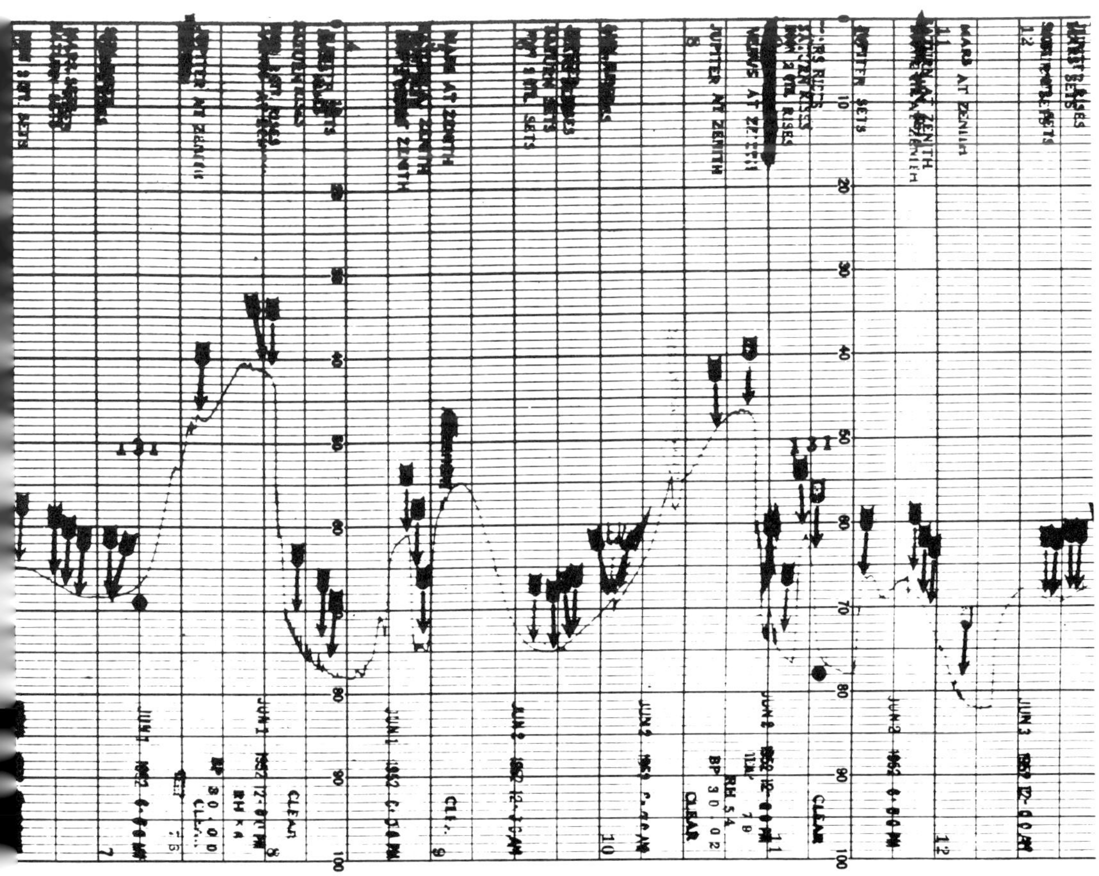

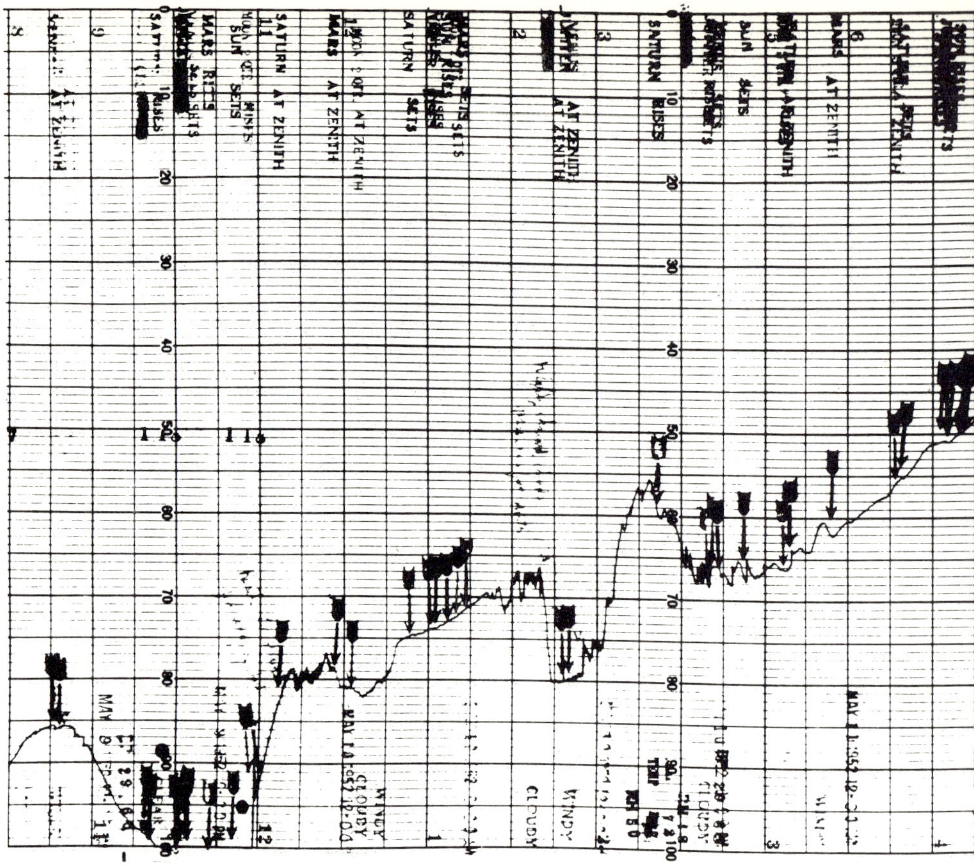

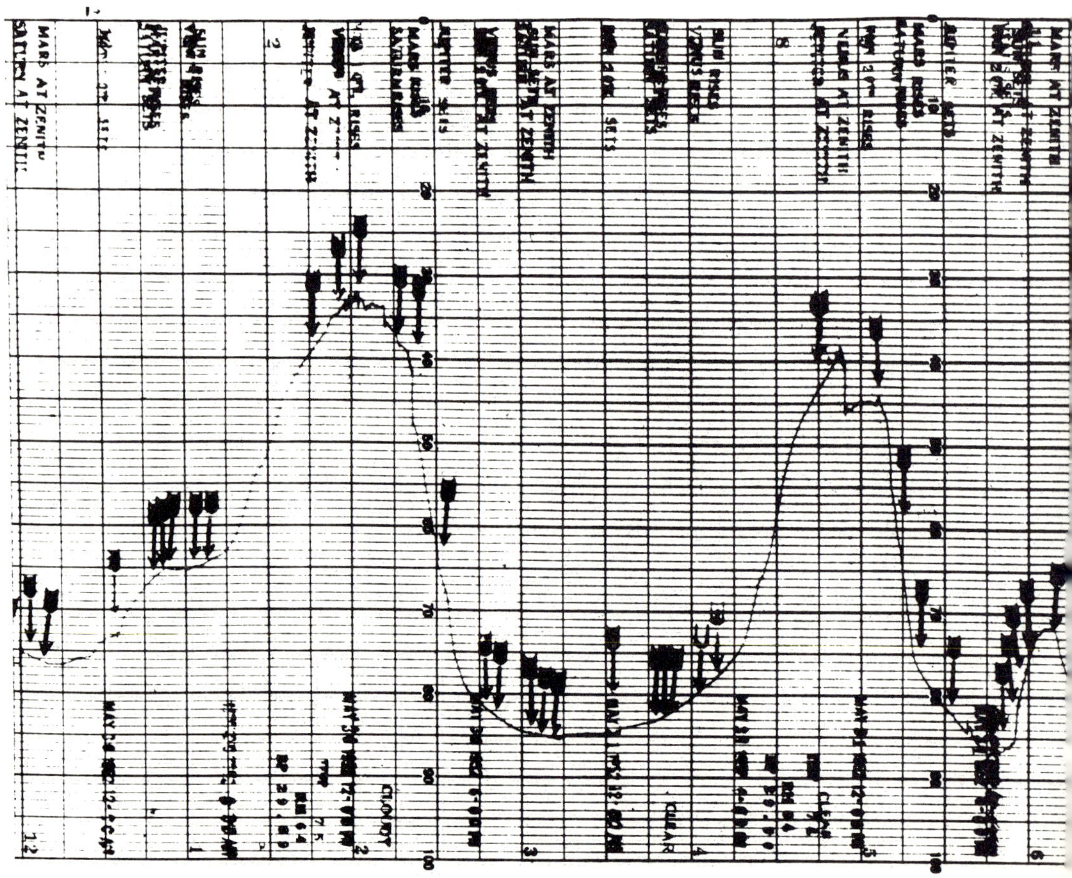

Fig. 158.

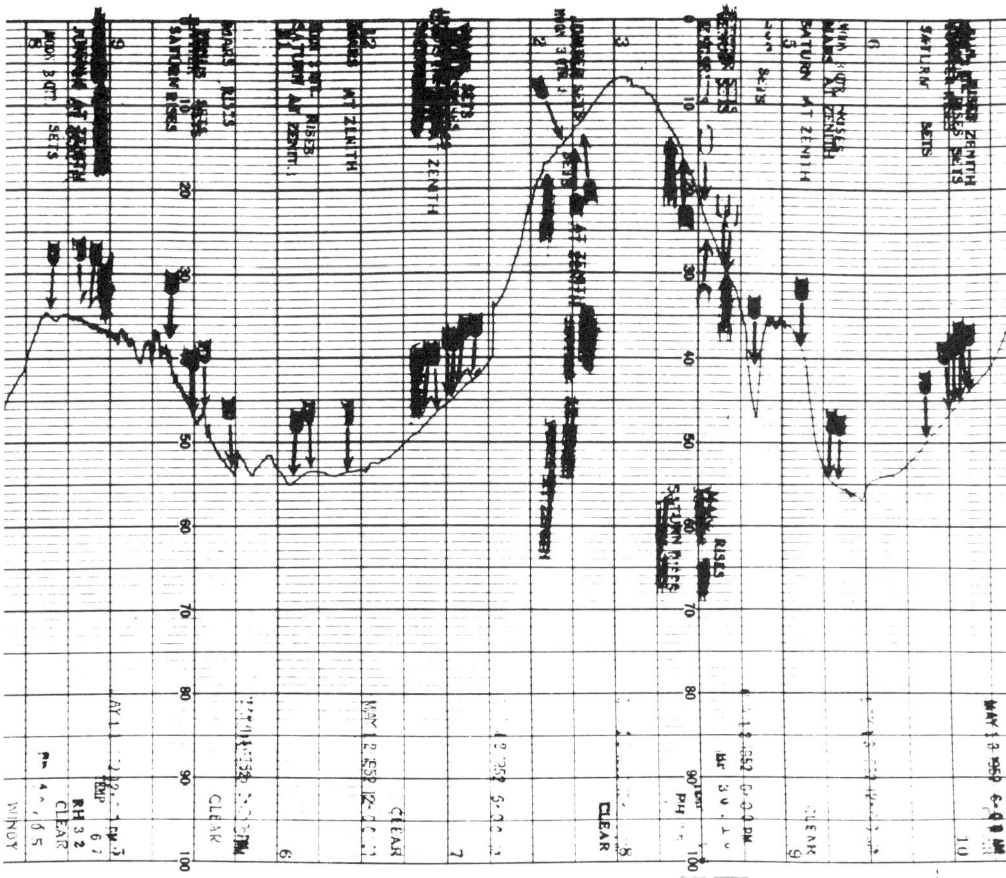

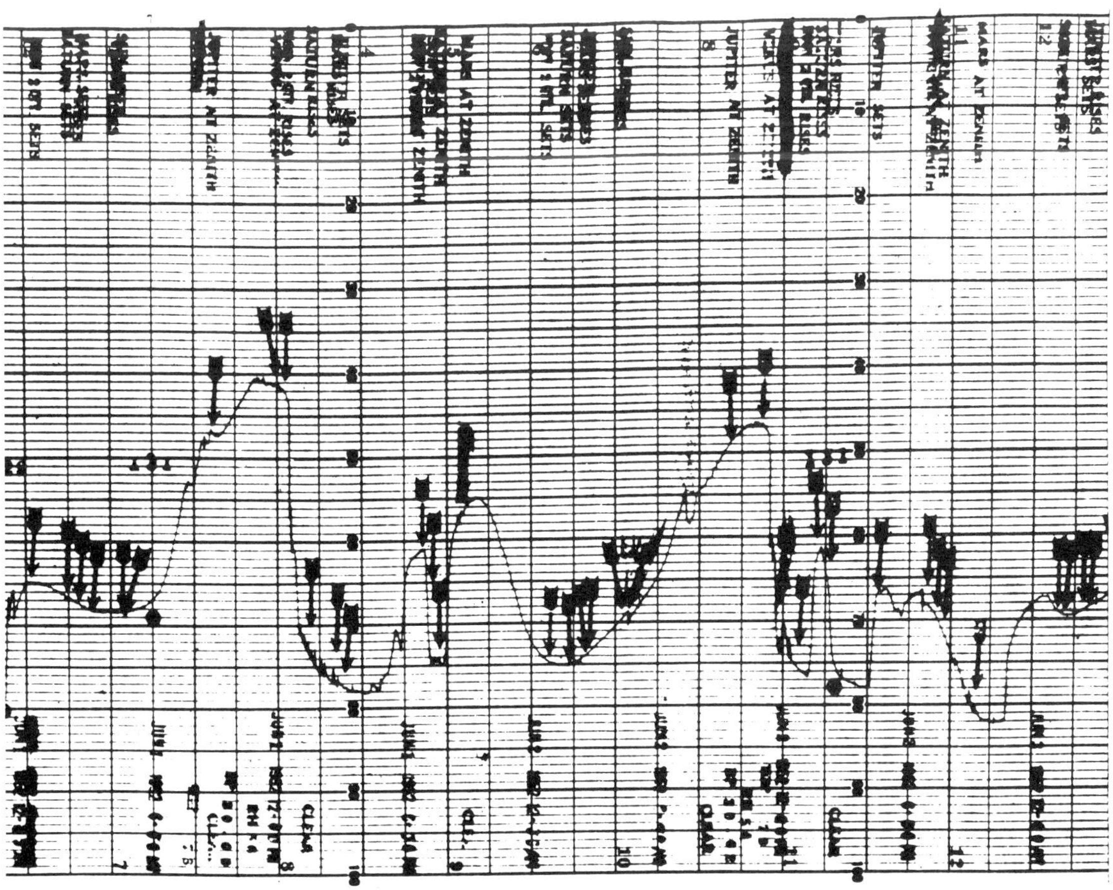

Fig. 159.

Fig. 160.

AUTOMATIC RECORDING EQUIPMENT

Fig. 161.

The Time of the Great Dying

Fig. 162.

Chapter 9

THE TIME OF THE GREAT DYING

The passing star caused the Earth to leave its orbit a distance of approximately twenty million miles. Although "the earth was removed exceedingly from its foundation" it returned to its original orbit within a short time. As the Moon was torn off, the Earth's poles were in the throes of inversion and had inverted about 90 degrees at this point. During the interval of this deviation a vacuum was created at the pole facing our Sun, as air and water were pulled to the equator by the passing star.

Then, as the star moved on into space, and its attraction lessened, the remaining Earth's atmosphere returned. Ice formed over the polar areas very swiftly. This action trapped many animals, preserving them in their original state, with flowers and grasses still in their mouths. Some of the animals were found in such perfect condition that their flesh "looked as fresh as well-frozen beef."

To date, scientists have been unable to explain the mystery of the bones and frozen animals in the far northern latitudes lying in and on the surface. Byron C. Nelson states in his book, "The Deluge Story in Stone,"* on page 125:

> "The remains of mammoths are incredibly numerous in Siberia and, strangely their numbers increase farther north toward the Arctic Ocean. Their bones are spread over the bottom of that ocean where ships have dredged them up. And 200 miles to the north, in the New Siberian Islands, not much farther from the North Pole than New York is from Chicago, mammoth remains are the thickest of all."

During the gigantic tidal flood of waters that rolled over the Earth for weeks prior to inversion, all life on Earth was destroyed and the great tide deposited great quantities of it at the edges of the tide which tapered off at the pole.

Some giant mammoths in the far north have been found still standing on their feet, their stomachs full of food, composed of sedges, buttercups and grasses. These plants were found freshly cropped in their mouths. The elephant's two-feet-long hair was still in its natural condition and undamaged. These mammoths, along with thousands of other types of animals, were found frozen in muck, a mixture of loose soils, gravels and clay, broken twigs, shattered trees and leaves. None of these animals were found encased in ice only.

The vacuum that occurred when the atmosphere was pulled away, quickly took away from these animals the right to breathe and they died without violence. The terrible cold that swiftly set in when

* Byron C. Nelson, THE DELUGE STORY IN STONE, (Augsburg Publishing House - 1931)

the atmosphere returned, froze the animals where they stood. Proof of this was the condition of their lungs, the masses of clotted blood in their chests showing major internal bleeding (due to sudden vacuum acting upon the blood vessels in the lungs), position of their eyes and their ruptured stomachs.

The cause of this sudden drop in temperature was the radical change in the atmosphere in this region which normally had been more humid and spring-like. The returning air was without moisture and therefore the high energies of the upper atmosphere were unable to ground normally, leaving the area as we find it today, dry and cold.

With the Earth tilted at its new angle (23½ degrees), the angles of solar radiation upon the atmosphere were radically changed. Sufficient heat, therefore, could not be generated at the surface of the Earth to create the required humidity, thus energies descending to the ground brought cold air down continuously from the upper layers of the atmosphere.*

Probably 75% of the remaining water on the earth was converted into cloud vapors by the terrific heat coming from the Pacific cavity and from the thousands of faults all over the remaining half of the crust. The great volumes of water and atmosphere that had been pulled away from Mars had volatilized in deep space and began arriving on Earth under the pull of Earth's gravity. The water vapor arrived in great volumes and as it gently cascaded down through the atmosphere, it formed into raindrops. It must have rained for many weeks before all the vapor was gleaned from outer space. Upon landing on the still hot, newly-formed ocean basins, the rain turned instantly to steam which rose again into the atmosphere. This process went on continuously for many weeks† until the ocean floors cooled sufficiently to allow the water to remain and build into the great oceans we know today.

Snow began to fall in the polar regions and also in the temperate zones for the first time. Before the catastrophe, snow was an unknown phenomenon on the Earth.

In the far north, surface waters began to drain toward the Arctic Ocean from the slopes of the newly-created hills and mountains all along the northern coasts of Siberia and Alaska as well as northern Canada. These waters carried with them great quantities of surface debris. When the torrential floods arrived on the coastal plains, they slowed and deposited their load of silt and rubble. Animals caught in the flood were covered and buried deep before decomposition could begin.

Close behind the floods came the cold winds and snows – destined to remain until the next catastrophe. This sudden lowering of temperature assured the preservation of the buried life. Many types of animals have been found in these morasses of death: giant tigers, giant bisons, wolves, beaver – to name a few. Many of these animals are now extinct.

The Siberian morass of death is only one of many. The deposits found in Nebraska, the Dakotas, California, Florida, Texas, and Africa, contain remains of the giants. Others will be found across northern Canada and Greenland.

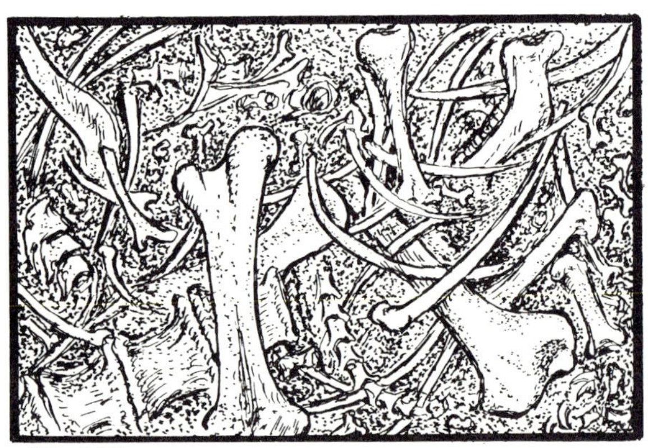
Fig. 163.

* See page 45 " The Primal Energy Transverter" by Lewis A. Manson (Copyright - 1966)

† The forty days and nights of rain and great flood of the Bible can now be readily understood.

Fig. 164. These buttercups were gathered in northern Siberia as small plants but grew to their present large size when transplanted to Southern California — the same size their ancestors were when found in the mouths of frozen mammoths!

Fossils of plants have been found at both the North and South Poles. The plants are of a type that grew in a mild climate and required sunlight – every day and all year – for growth. Although the climate in these areas has changed considerably, these continents have also moved hundreds of miles from their original position.

Some of the plants found in the mouths of the mammoths still exist in southern Siberia, but are extinct or found in "minature" along the northern coast. This clearly establishes the climatic change that occurred along with the crustal movement.

Most of the land animals perished as fissures opened nearby or beneath them. They were swept into deep chasms by the sheets of water rushing over the surface. The surviving dinosaurs, flying reptiles and other animals of monstrous size were left stranded on comparatively dry land with no source of food remaining since they had lived upon the abundant plant and animal life in the huge shallow seas that existed no longer.

These giants of the past, heavy and cumbersome, were completely unequipped for the new environment beginning to form around them. They could not adapt, so died in a very short time. Some of the smaller animals and man were able to adapt to the new surroundings and so survive to this day.

Fig. 165. A nest of fossilized dinosaur eggs.

Petroleum was forming just a few days after the star entered the solar system. The increasing intensity of gravitational attraction caused the Earth's crust to fracture early and to progressively

*See THE CREATION OF PETROLEUM in this series.

shift, distort and swallow great volumes of water. The seas poured into the fractures continuously, carrying with the great volumes of water enormous quantities of dead and dying marine life along with plants of every description.

The entrapment of this life at the bottom of the millions of deep, gigantic chasms all over the Earth, set the stage for the rapid creation of petroleum. The heat thirty to sixty miles down (where these life forms were trapped) was several hundred degrees F, and drove the water off as steam. At the same time, the surface soils were sloughing into these chasms, mixing with and covering the organic debris – thus sealing the debris deep in the Earth.

Later, after the great turmoil ended and the faults had slammed shut compressing the entombed material, the heats in the Earth, acting like a catalytic cracker in a refinery, began the process under great pressures of converting the material into the hydrocarbons we know as petroleum.

The generated pressures drove the newly created volatiles up into porous formations near the surface of the Earth to be used by man at a later date.

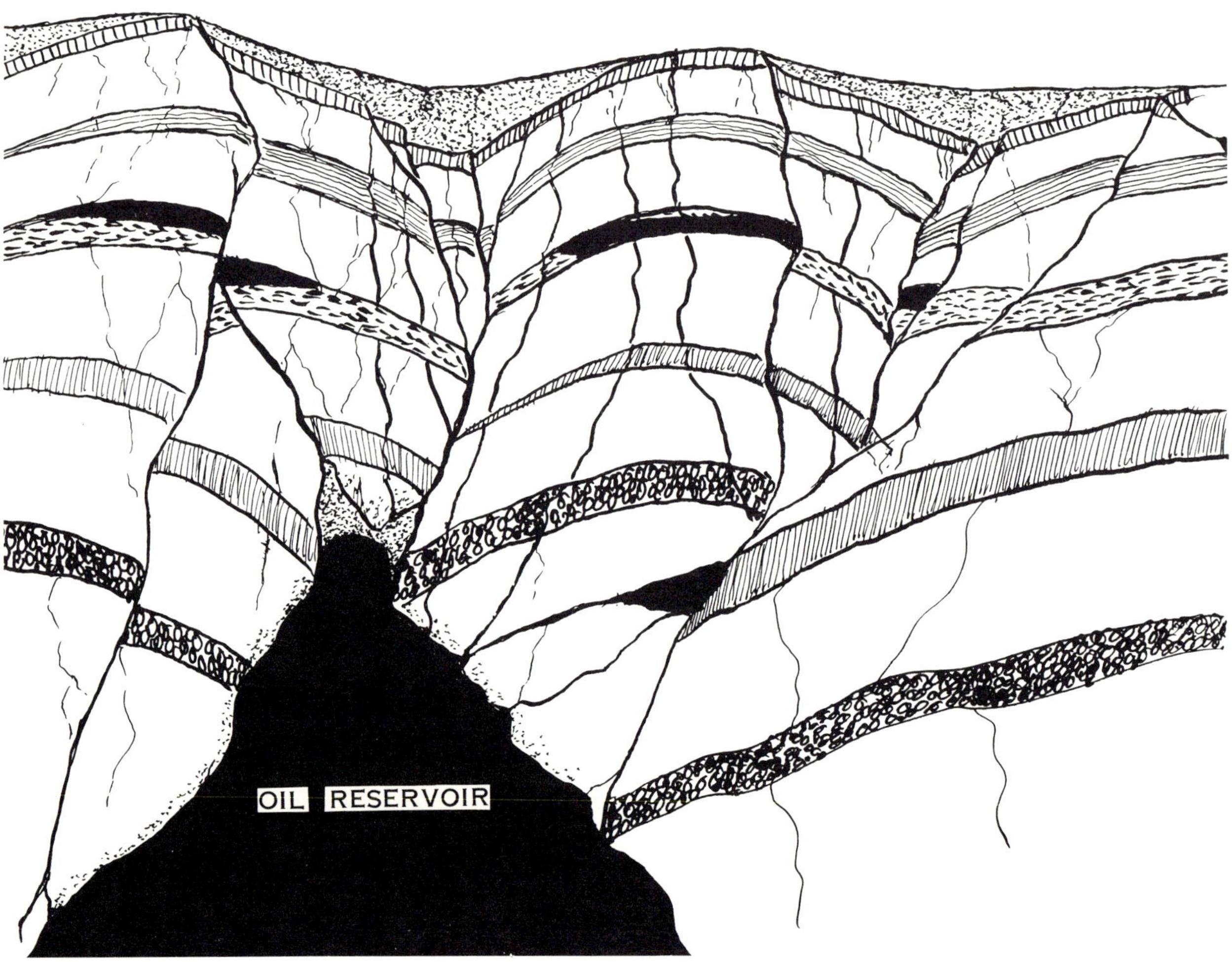

Fig. 166.

The Formation of the Ocean Basins

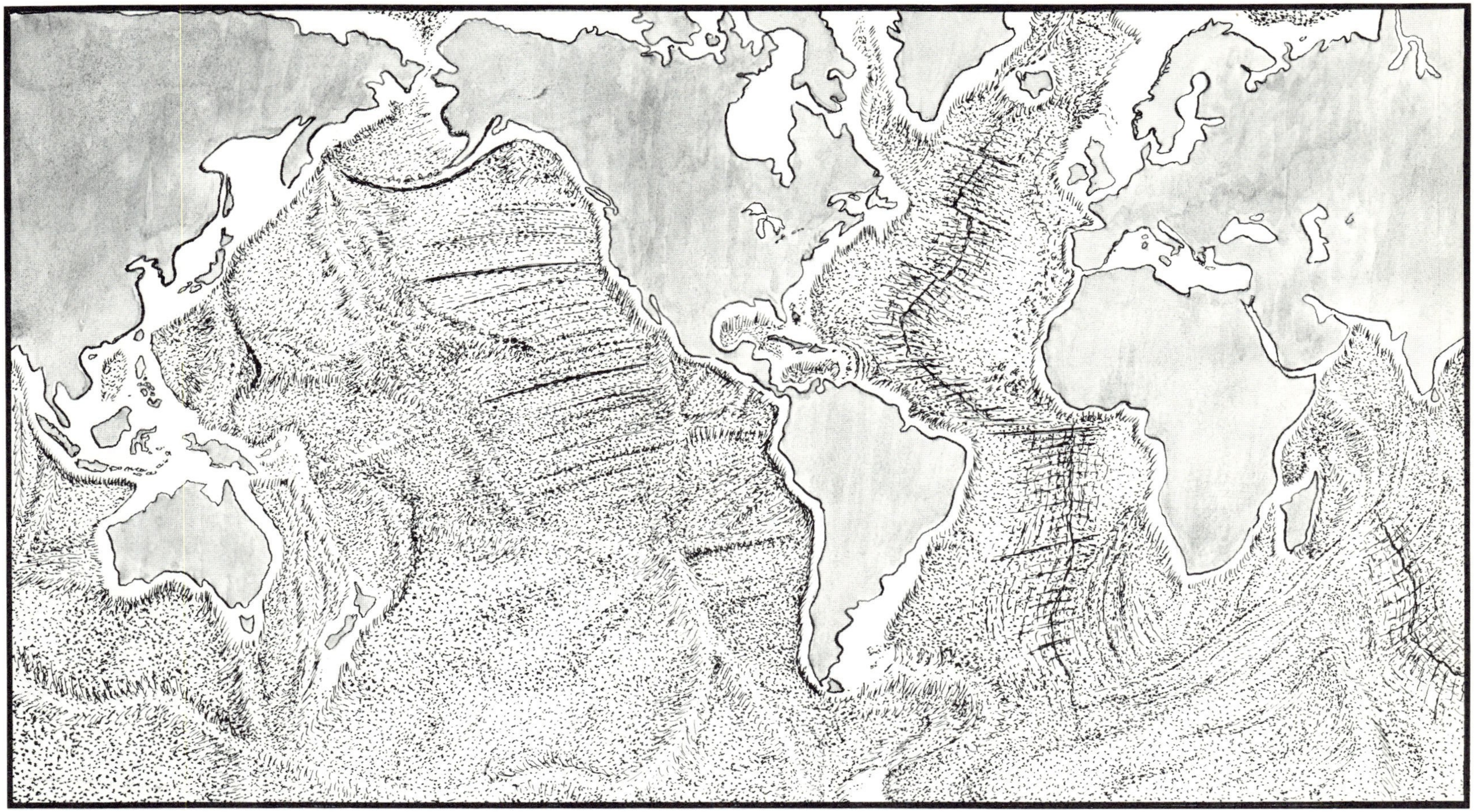

THE TOPOGRAPHY OF THE OCEAN FLOORS

Fig. 167/168. Shaded areas are the continents. White areas are the continental shelves. Stippled areas are the ocean basins.

Chapter 10

THE FORMATION OF THE OCEAN BASINS *

The topography of the ocean floors reveals a record of great significance. The long, unbroken mountain ranges, escarpments, fractures and deep trenches, plains and plateaus, the distinctive system of the integrated rises, especially the East Pacific Rise, indicate that the formation of these ocean basins could have come about in only one manner – the distention of the Earth's crust. The distention of the crust was also the true cause of the formation of the continents. There had to be vast areas of surface crust removed to allow for the great drift of the remaining surface crust.

First of all, beyond the continental shelves, which generally lie below sea level, the type of rock changes abruptly from sedimentary† to igneous.+ The materials of the ocean floors in general consist of basalts, granites, and numerous other igneous rocks.

In Fig. 169, the central point from which the Moon departed is located at the center of what is called the "East Pacific Rise." This is a subterranean mountain range. Oddly enough, Easter Island (the place of the strange stone statues all facing east – except seven which face northeast) is the center landmark from which the Moon left the Earth. The site being 110 degrees West Longitude, 63 degrees South Latitude.

The "navel of the world," or point of parturition, is located beneath Easter Island and is the focus of the Moon's extraction. The natives of Easter Island also call this area "The Navel of the World."

I suspect that a plug of the Third Vitreous Interbasement (12th layer) may have been removed along with a portion of the basalt-feldspar-orthoclase layer (13th). This plug would have been some 30 miles thick but of unknown diameter, although I do not believe it was very large. The crust that constituted the Moon was slipped off the Earth along with great quantities of liquid rock in a period not exceeding 6 hours.

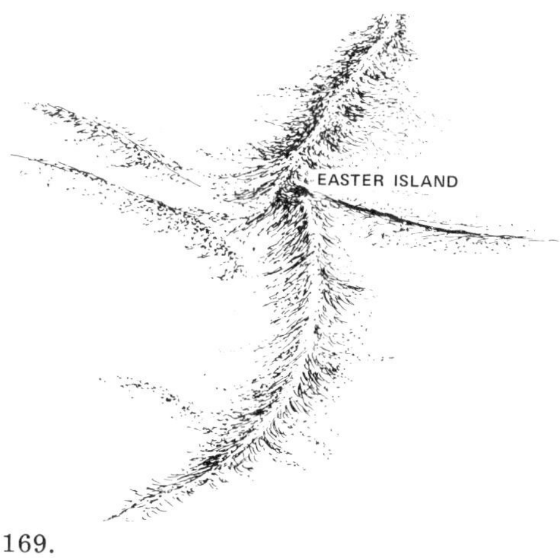

Fig. 169.

* This chapter is of the utmost importance since all other factors in this work are clearly confirmed by the evidence revealed on the ocean floors.

† Sedimentary rock (pre-catastrophe) is the surface material that arrived from outer space. This material is produced by the Sun and the stars and arrives on Earth at the rate of 1 inch every 5000 years. It is this material that life of all forms thrived in and upon, and left their fossil impressions within, all through time.

\+ Igneous rock is any material that became the crust at the time of the Earth's creation .

The moment the Earth's crust left, the Third Vitreous layer (Fig. 32) began fracturing in many places nearest (at the point of opposition) to the passing star. Only that area on the ocean floor which was in opposition contains this evidence. The great fractures which are still partially visible on the floor of the Pacific Ocean are the Marquesas fracture zone, a fracture just north of Sala-y-Gomez Ridge, Clipperton fracture zone, Clarion fracture zone, another at 135 degrees west to 145 degrees west longitude and 65 degrees north latitude, the Murray fracture zone, Demendo Cendo fracture and the Cascadia fracture. The true lengths of these fractures have been obscured by liquid rock fill. There are many other minor fractures which are not visible.

These fractures and the arc line of the fractures tell us much about this great event. First, they tell us which portion of the Earth was facing the passing star. Next, the length of each fracture tells us at which point the greatest attraction of the passing occurred. Thirdly, it shows that the Earth was inverting during the catastrophic event (Fig. 60). Fourthly, it tells us what fluid layer was contributing to the mass of the Moon since other materials beyond the range of the Pacific Rise are different and arrived immediately after the materials that made up the Pacific Rise. Fifthly, even the hour that the great forces began wrecking the Earth and the duration of the highest intensity of the forces was made evident in the area where the Moon departed. Sixthly, it even tells us when the first winter began on the Earth since inversion terminated at the new "North" Pole, but left the Earth with a tilt of 23½ degrees away from the Sun. Lastly, it tells us in which sector of the planetary orbits the star passed.

Now – let us examine each item carefully:

1. We know that the seven major fractures beginning at Antarctica and ending near Alaska are the only kind and type of fracture series existing anywhere on the Earth. We also know that only fractures of this immense size could originate only from a very thick and quite vitreous layer, which is the Third Vitreous layer (Fig. 32). Only tremendous forces could fracture this layer.

In fact, were it not for this arc of fractures in the eastern Pacific – earthquakes would be of much greater intensity in this area, due to the great pressure surges of liquid metals continuously rolling below the Third Vitreous layer as they do in the western Pacific under the New Hebrides, Japan, Aleutians and others. However, due to these fractures, the pressure surges lose much of their force as they pass below these great fracture zones, since the fractures expand under pressure and absorb much of the shock generated by the surges.

2. Although only seven major fracture zones have been discovered, there are several more that remain to be found. These fracture zones began under the area that is now the Antarctic continent (the fracture lengths here are rather short) and increase in length as they approach Pitcairn Island. Beyond Pitcairn the two greatest fractures are the Marquesas and Clipperton fracture zones. These are of the greatest length, although much of them are obscured by the masses of materials constituting the East Pacific Rise. It is in this vicinity where the greatest attraction by the passing star played upon the Earth's surface and to which great surges of the liquid metals in the other layers deep in the crust pressed against the base of the Third Vitreous layer causing it to

fracture. Had the gravitational intensity of the passing star been much greater, a larger section of the Third Vitreous layer and portions of the Fourth Inter-metallic layers would have become a part of the Moon, resulting in an Earth of much smaller dimension.

However, since the Earth was inverting swiftly and at the same time rotating, the deep pressure surges were constantly shifting to new positions deep in the Earth's crust, thereby preventing greater disaster to the Earth. At the instant the Moon left, the Earth had inverted* 50% (the present North Pole was pointing to the Sun and the South Pole to the passing star) the gravitational attraction then began waning which resulted in lesser fracturing of the crust, as shown by the steady reduction in length of the fracture north of the Clipperton fractures (Fig. 171).

3. After full inversion of the Earth was achieved, the Sun stopped further tumbling by taking over control of Earth's flight in space. The curve (arc) of the path of fractures traces the inversion episode.

4. The material of the Pacific Rise is made up of former liquid rock that existed in the 14th layer. The balance of the material in the Pacific Ocean floor is composed of the 13th liquid rock layer (basalt, feldspar, orthoclase) and many metals ejected through fractures in the Third Vitreous layer.

5. The fracturing of the Third Vitreous layer began at 75 degrees west longitude under what is now the Antarctic continent and gradually curved over to 100 degrees west longitude, 25 degrees south latitude, which was the point at which the Moon was extracted from Earth as inversion progressed and rotation of the Earth continued. The fracturing pattern continued past the 13th west longitude to the 150th west longitude and ceased at the 165th west longitude at 75th north latitude (Fig. 171). In all, the total time consumed for full inversion of the Earth and extraction of the crust required six to eight hours.

North of this fracture, the fluid rocks beneath the crust drained north into the newly formed Arctic Ocean. All crust north of this fracture dips north. South of this fracture the downwarping of the crust created mountainous terrain for thousands of miles. Then southward the crust dips toward the Indian and Pacific Oceans. Many earthquakes occur along this fracture.

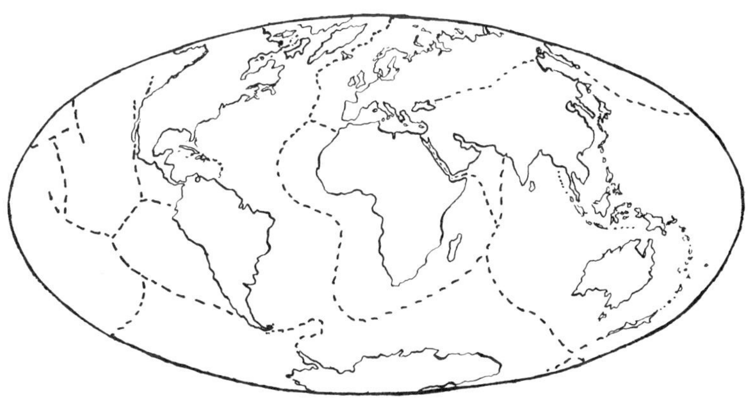

Fig. 170.

* Venus and Mercury are the only two planets that were not inverted. The Sun still rises in the west on these planets and sets in the east, which follows, therefore, that magnetic north on these planets is opposite that of the other planets. However, the magnetic field of Venus may present a paradox due to the arrangement and volume ratio of its layers.

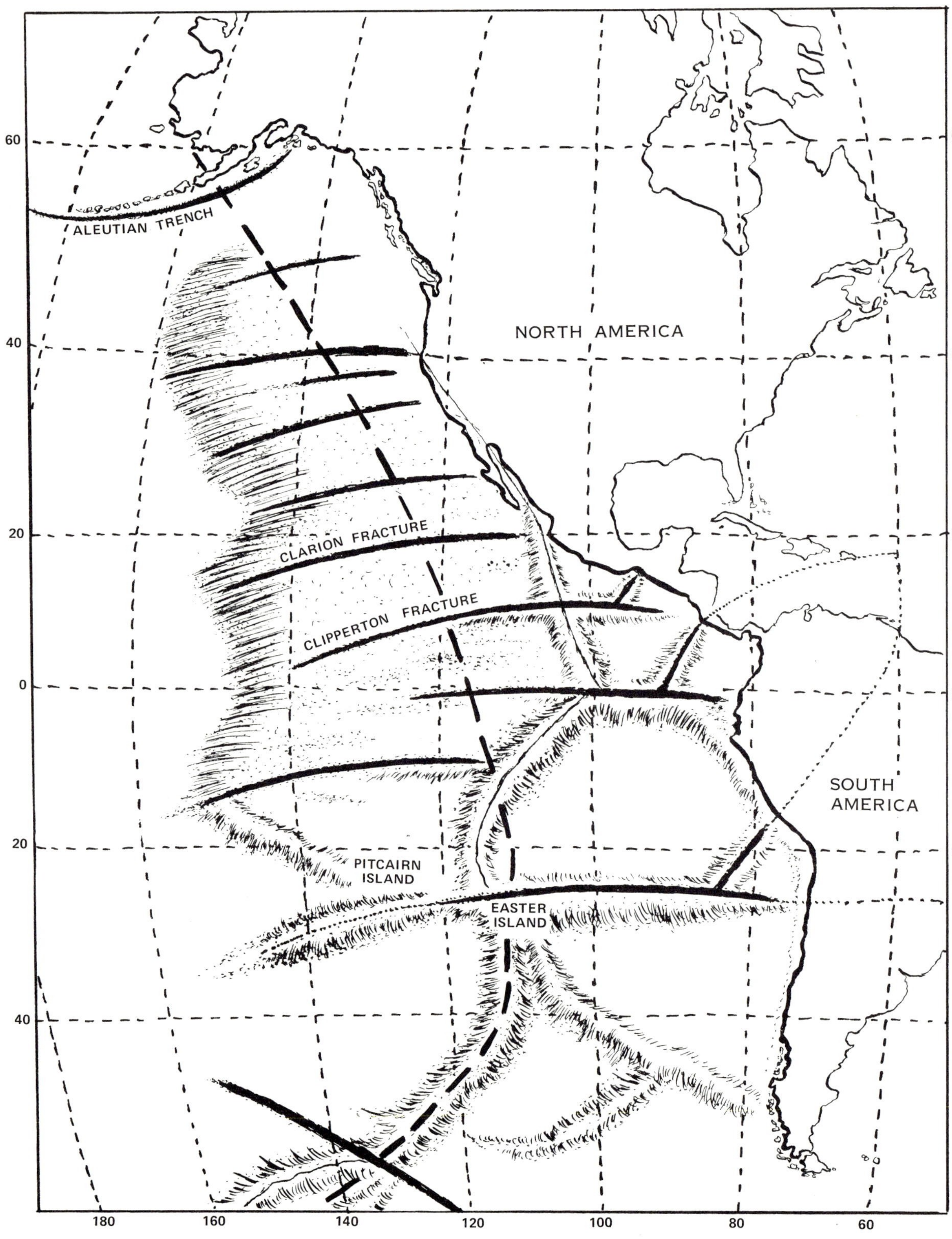

Fig. 171. NOTE: The arc of the fractures show the track of inversion.

The same moment that the Earth's crust began mounting into a heap, as it slipped over the Earth's surface to depart and form into the Moon, the first major continental fractures appeared in the remaining crust. The great gravitational attraction of the star began to swiftly move the Antarctic continent, South America, North America, Australia, and many other minor masses along with the Moon-forming crust. In fact, had the great gravitational attraction been sustained but a few minutes longer, parts of central South America, Antarctica and Australia would also have become satellites of the Earth, but as it was, the gravitational attraction began diminishing and with the inversion of the Earth in progress, the position of the continents approaching the danger zone were moved out of jeopardy.

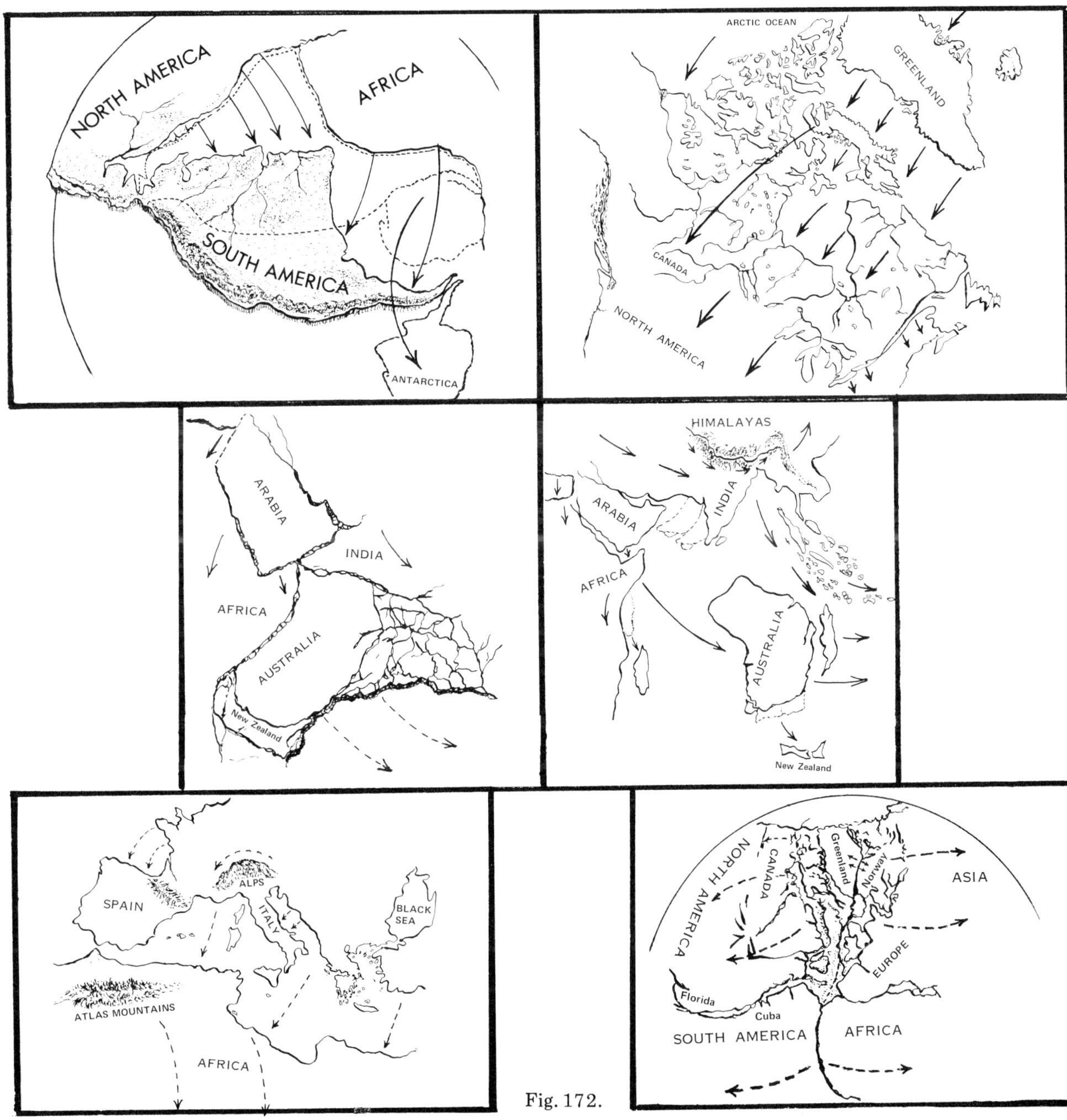

Fig. 172.

The next series of events to take place was a continuation of great turmoil and unimaginable chaos. In the western Pacific, many long crustal sections fractured away from the Asian coastline and swiftly rode out atop extremely hot liquid rock as it rushed from beneath the remaining crust out into the newly formed Pacific basin.The Aleutian chain is a long string of crust that drifted a few hundred miles and stopped early because the fluid rock became depleted and allowed this section of crust to settle upon the hard surface of the Third Vitreous layer.

Over in the Atlantic basin is a great undersea "mountain" range extending from Antarctica to Iceland. It is called the Atlantic Ridge (Fig.168). In the North Atlantic this ridge is quite broad and was created as the North American continent broke away from Europe. At the moment this great fracture appeared, the 14th layer of the crust welled up to fill the fracture. However, the fracture continued to widen, and as it did so, the liquid materials poured out from beneath the crust of Europe and North America. This rock hardened quickly in the naked atmosphere at the surface almost as fast as it became exposed.

Gradually the elevation of the ridge on each side of the fracture's center became lower as pressure and supply of extruding materials dwindled, but the North American continent continued to drift with liquids pouring out from beneath it into the Pacific basin and Atlantic basin, but not as violently into the Atlantic as the Pacific. This latter stage of the flow created the flat planes of the floor of the Atlantic such as the Sohm Abysmal Plain. Great flows of water helped chill and harden the liquid rock as it became exposed. This allowed it to retain its present form. Water was rushing over the Earth's crust while hot liquid rock flowed under it.

As South America swung in an arc Fig. 173)* the same activity took place, but since this portion of the crust moved in an arc, the breadth of the Atlantic ridge was quite narrow and the altitude of the ridge above the floor was considerably less. This area of the ridge lies on the equator between 15 degrees west and 30 degrees west longitude. However, the balance of the ridge to the south again widens due to the westward movement of South America which began after the eastern tip of Brazil had passed below 10 degrees north latitude.

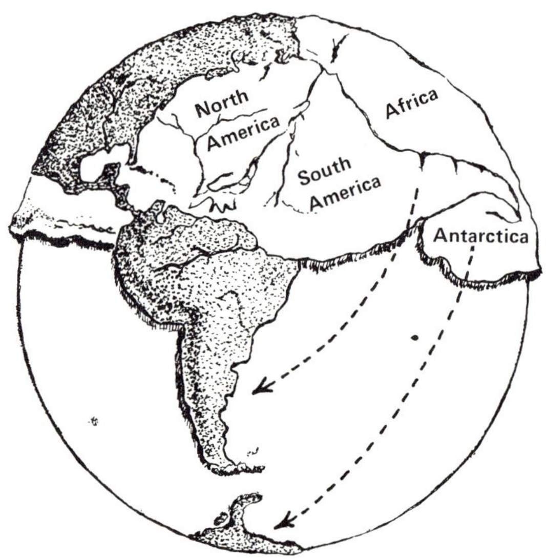

Fig. 173.

* The greatest error in the Continental Drift theory is the assumption that the northeast coast of South America parted from the west coast of Africa. Actually, the north coast of South America parted from the east coast of North America.

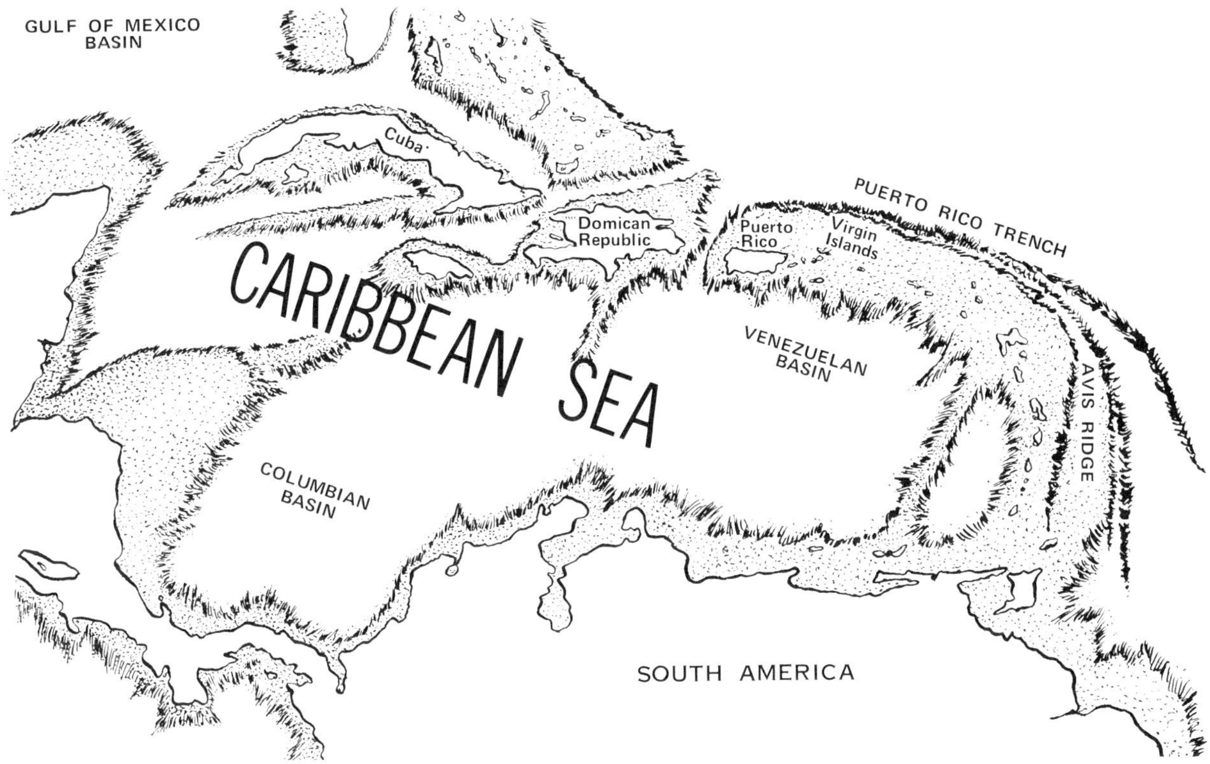

Fig. 174. Note the tell-tale arc of the Avis Ridge left by the north coast of South America as it swung in an arc downward after breaking away from the east coast of North America.

A great wave of liquid rock was set up as South America moved southward. This wave was on a line east and west on the 20th south latitude and is called the Trinidad Seamount line. A very interesting feature of the arc movement of South America is seen at the mouth of the Puerto Rico Trench (Fig. 174). From that point, two deep fractures curve from the east southward several hundred miles indicating the change in direction of this great continent's movements.

The ocean floor east of the Atlantic Ridge is generally shallower than the floor west of the ridge. The reason for this is quite plain since it took a much greater volume of liquid rock to fill the Pacific basin, with the greater flow of liquid rock moving west with the continents. The Walvis Ridge and the Guinea Ridge were formed as a result of the eastward movement of South Africa (this movement was not much more than a hundred miles).

A few crustal remnants that drifted into the ocean basins, either trailing the Moon-crust or a continental block, were Iceland, Cape Verde Islands, Bermuda Island, Antilles, Fiji Islands, New Zealand, New Hebrides, Formosa and many others.

Had the Gulf of Mexico and the Caribbean Sea not been blocked by the string of islands from Cuba to Trinidad left by South America as it swung into its present position, the depth of these basins would have been equal to the Demerara Abyssal plain. Evidence of the flow stoppage is due to two great waves of liquid rock which formed. The first, Avis Ridge, and the second, Beata Ridge. The same is true of the Mediterranean Sea, although this flow was stopped as Spain slammed into the coast of Africa.

In Fig. 168 is a view of the North Polar region. Here may be seen how the North American continent, including the lands of Northern Canada, Baffin Islands and Greenland, drifted away from the European and Asian coast. Here again, three ridges were created – each 90 degrees to the direction of the drift of these land masses.

In the vicinity of each great fracture in the ocean basins, great volumes of metals poured out under fantastic pressure and were mixed with the surface liquids as they were flowing to balance the Earth. Very soon after this, additional readjustments caused intermittent expulsion of metals to splatter over the floors of these basins which were beginning to harden at the surface. This is why nodules and nuggets of numerous metals are now found strewn over the ocean floors.

Fig. 175.

The western Pacific is heavily scattered with islands, chains of islands, long strings of land masses and long narrow crustal segments (Fig.175). These are the initial portions of the Asian continent which moved with the greatest velocity eastward (in an attempt to leave with the Moon) their direction of movement was generally toward the northeast – the direction the crust left the Earth on the eastern side of the Pacific Ocean.

There are only a few scattered islands and other crustal segments off the North American continent. These were set adrift from this continent and rode out to the west to meet the liquid materials arriving from the western side of the Pacific. The attraction of the passing star helped prevent the creation of large numbers of islands leaving the west coast of North and South America. Most of the Pacific Islands originated from Australian, African and Asian coasts. You will note, too, that off the coast of America all the way to the East Pacific Rise there are no islands to speak of due to the arc movement of South America into position since the Antarctic continent carried these islands with it into the South. You will also note that few islands exist in the Atlantic basin due to a much less violent separation of the crust.

The wounds of the Earth and the Moon healed quickly – but life remaining on Earth began a long arduous task of adapting to an ever-changing environment on a completely changed world.

So ended the Cretaceous Period of the Zenizoic Era and the Neozoic Era began.

Fig. 176. Remnants of the birth of the Moon off the coast of Oregon.

Fig. 177. Additional fragments of the crust left stranded off the Oregon coast after the birth of the Moon.

The Great Ocean Transfer

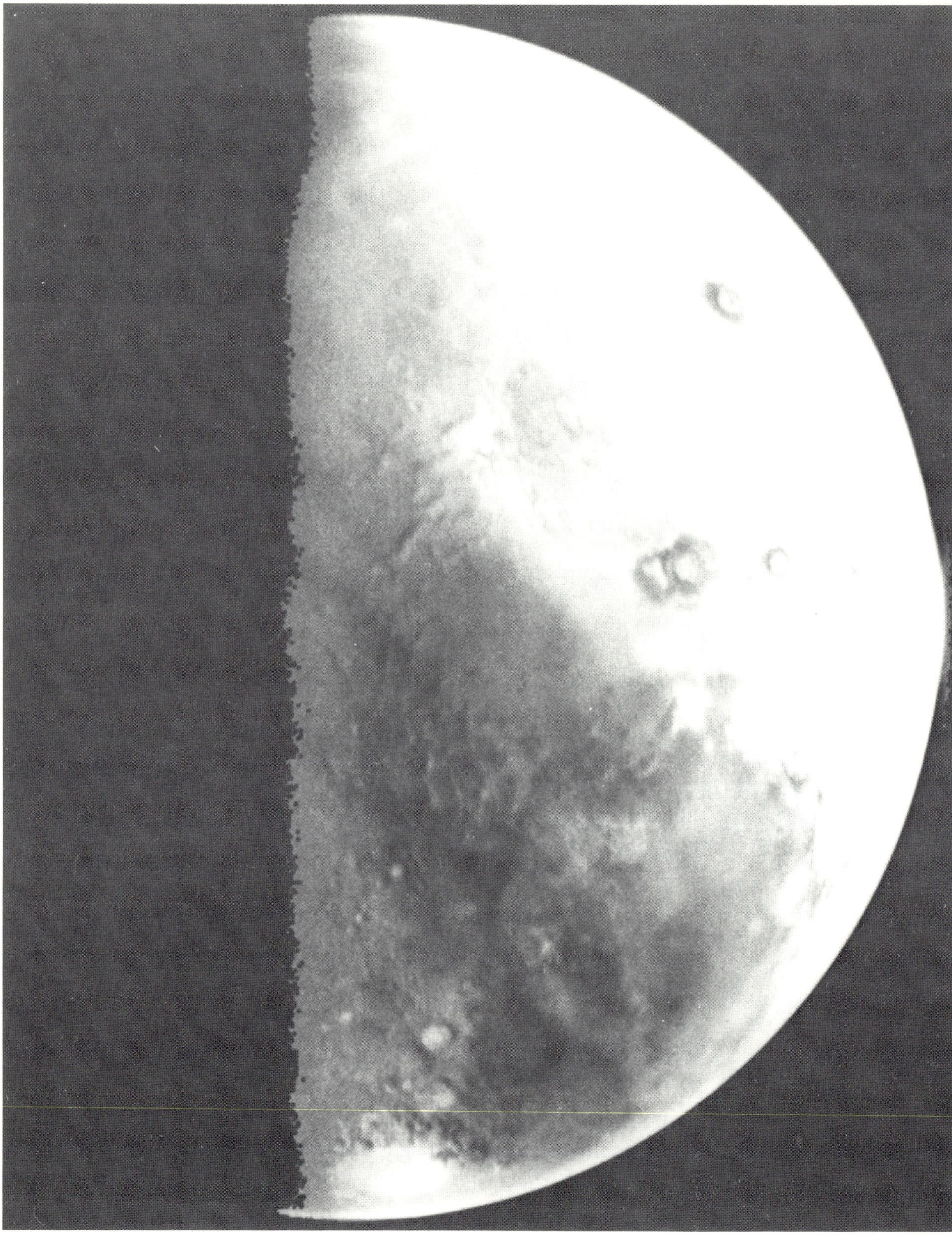

Fig. 178. MARS NASA

Chapter 11

THE GREAT OCEAN TRANSFER

Much pondering and debating was done covering the subjects of this chapter and serious considerations were made concerning the advisability of even including the subject of the ocean transfer in this work, since it is almost in the realm of fiction rather than reality. However, since many other occurences in the past have bordered on the fantastic or unbelievable, it was decided to go ahead and include this chapter in the book.

I wish to remind the reader of this work that the subjects which it covers ranges from the smallest parts of the atom through not just our galaxy but the total universe, and the reader must bear in mind at all times the scope of these subjects. Many of the subjects discussed in this work are things of enormous size and we tend to minimize the size of planets, suns, stars or universes with relation to the size of the universe. A few "droplets" extracted from the Sun's surface is actually a very small activity requiring a very short period of time. The little balls of fire, called stars, out in space and the covering on them is actually very thin and the material extracted from them by forces almost incomprehensible in scope were done with great ease. Therefore, the activity discussed here is not beyond the realm of possibility by any means, but rather a simple activity taking place under natural causes. The reasoning presented here is based on a natural course of events during this activity. *The sequence of events described for the birth of the Moon and the destruction of the Earth's crust occurred on Mars in like fashion.*

As the intruding star approached below the plane of the orbits of the planets, it began to pull away the atmosphere of Mars and followed quickly by the waters in a great flood. The waters rose in volume beyond the point of a great tide and began to leave Mars in large quantities. As the atmosphere of Mars moved on out into deep space it expanded rapidly into a very thin mass which automatically slowed its departure and movement toward the star. A short time later the waters that left Mars and were now moving into deep space, were greatly affected by the great heats (radiation) from the star which tended to vaporize much of these waters. This vaporizing process also slowed the movement of the water toward the star by dispersing the great liquid globules into a large gaseous mass. However, not all of the water was vaporized.

A short time after the departure of the atmosphere and waters, Mars, being pulled down out of orbit, was forced to relinquish two great sections of its crust which became the two satellites, Phobos and Deimos, now orbiting around Mars. Where Mars formerly had a circular orbit, it now follows an elliptical orbit because of the activity and disturbance to its former course.

From several recent photographs of Mars taken by some of our space vehicles, the flow lines of water exiting the planet are evident.*

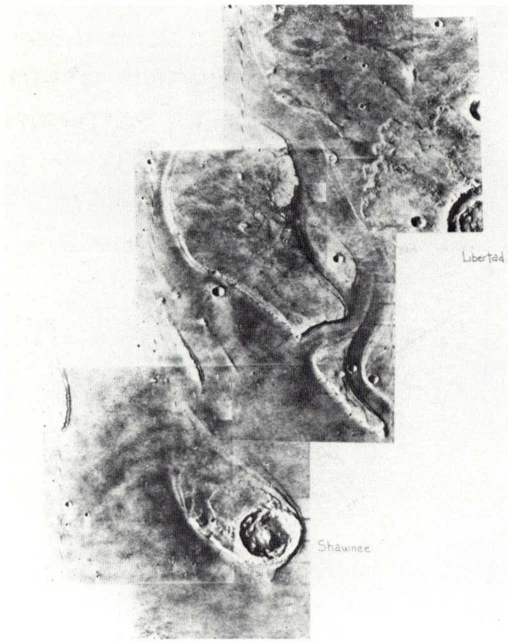

Fig. 179.

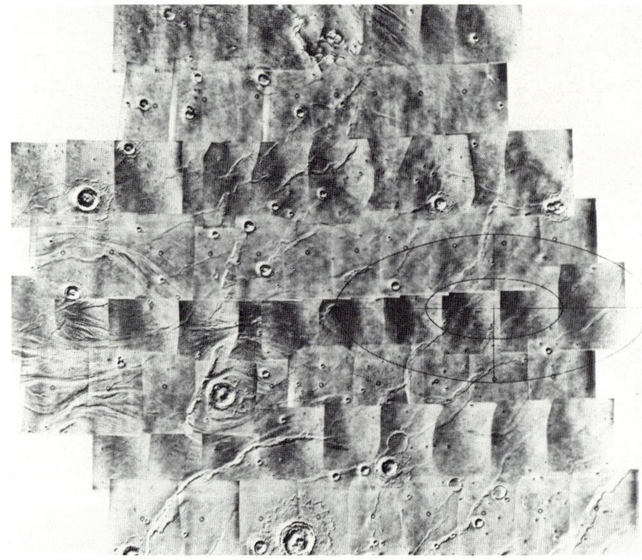

Fig. 180.

Surface of Mars showing flow lines of once soft mud. NASA

The surface features on Mars are the result of the muds flushed from the surface as the waters left the planet. These muds are the residue of the great masses of water that left, and since the star was at a much greater distance from the orbit of Mars, the mud masses were unable to accelerate fast enough to follow the wake of the star. They were lost in space at a distance where no other body in the solar system could attract them, except Mars, which pulled them back to its surface and thereby created the great numbers of craters on Mars' surface.

As the star passed Mars, its attraction began to effect Earth more and more. It began to pull away the atmosphere of the Earth and the waters of the Earth quickly followed (told in detail in Chapter 3). Now, since the Earth was in closer proximity to the star, the atmosphere and waters of the Earth were pulled into the wake of the star and followed it for many millions of miles until they,too, were left stranded in deep space, far from the Earth's attraction. The star passed on ahead of the planet Saturn and beyond, causing the planet to give up nine sections of its crust. One retrograde and the other eight in the opposite direction. Immediately after the star passed Saturn, the atmosphere and waters of the Earth were captured by Saturn and became a part of its rings that now orbit that planet. Not all of the lost atmosphere of Earth was captured by Saturn. Some of it was picked up by Jupiter as it passed by in its orbit. Venus also gained part of the Earth's atmosphere.

*** The Bible, and many other ancient books, mention the great flood and portions of the catastrophe. Books , such as the Bdok of Mormon, the Book of Jasher, for instance, not only mention the flood, but as in the Bible also mentions the 40 days and 40 nights of rain. These are two different episodes. One is a flood (caused by gravity) on which the ark rode the face of the waters for many days; the other is a deluge of rains that lasted 40 days and 40 nights.**

After the star passed on into space, the atmosphere of Mars began to arrive on Earth very rapidly — a new kind of atmosphere: more oxygen, hydrogen, less nitrogen and waters that were more brackish than the Earth originally possessed. It required approximately forty days and nights for the enormous volumes of water from Mars to land on the Earth. The initial atmospheres and water from Mars arrived in the direction of the *new* south, which is now in the direction of the Antarctic. The waters arriving on Earth that had not already vaporized did so now because of the passing star's great gravitational attractions. The waters traveled at low speed down through the Earth's new atmosphere. The waters arrived on Earth gently during the tug-of-war between the star and Earth, for if they had arrived at tremendous speeds they would have gouged great holes in the Earth's surface and these are not in evidence.

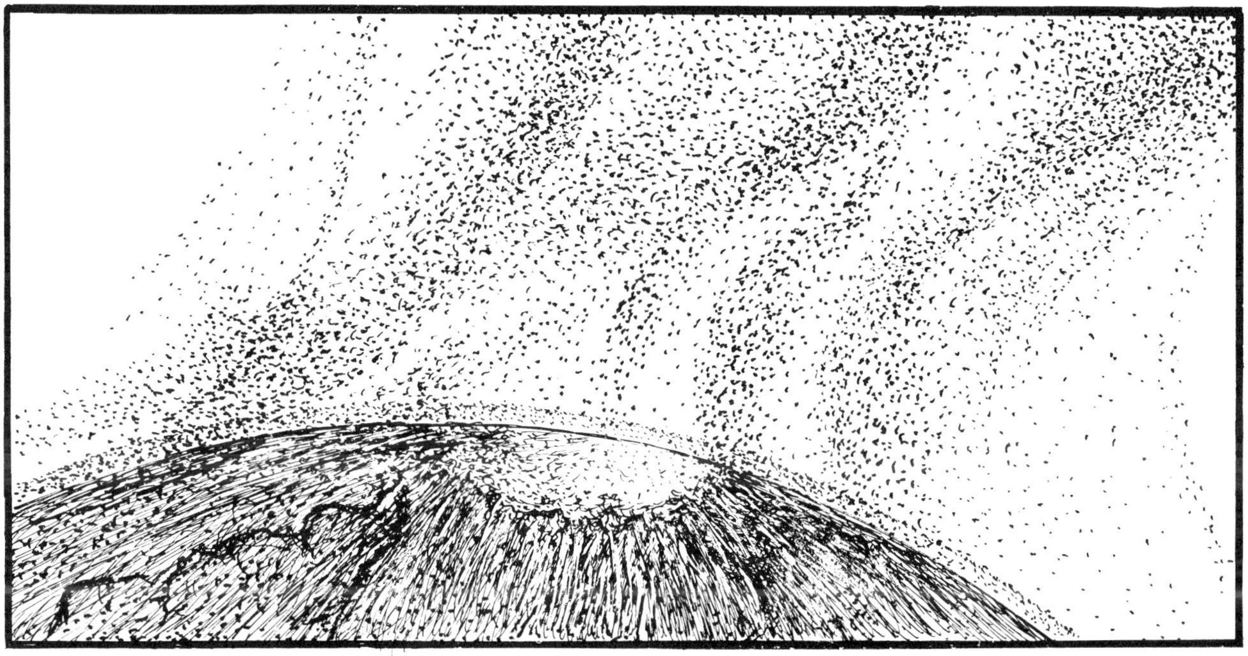

Fig. 181. The waters from Mars arrived on Earth gently.

Another enigma looms, at once important and curious. It has often been noted that there are many forms of marine and plant life on Earth today that did not exist prior to the catastrophe. After the catastrophe they suddenly appeared.

Thus, whole plants or their seeds or marine forms or their eggs were transferred from Mars to the Earth and survived the transit through space.* Since these life forms had been extracted from Mars alive or viable, they were protected to a degree in the muds or in the waters that left the planet and were not long in transit between Mars and the Earth.

There was a great change in the ratio of Earth's water from its previous depth (before the catastrophe) to its present depth. At one time in its history Mars had a great deal of water — far more than the Earth. Now it has none. If the oceans of the world — all the water on Earth — were lying the same depth over every square mile (that includes the ocean basins and the continents) they would cover it to an average depth of slightly over 6,000 feet. Whereas, prior to the catastrophe, the water depth on Earth averaged a little over 100 to 200 feet.

* Book of Mormon-Third Nephi, Chapter 8, Verse 16: "And there were some who were carried away in the whirlwind and whither they went no man knoweth say they know they were carried away."

Since a little over half of the Earth's thin, outer crust was removed and the remaining crust broke into segments and drifted into the present positions as continents, this left these continental blocks standing above the barren Third Vitreous inter-basement layer an average of 7,000 feet. Now, if the Earth prior to the catastrophe had been covered with water to a depth of 6,000 feet, the plants and animals of the Devonian, Pennsylvanian and other geologic eras, including the Cretaceous, could never have existed; since even a water cover of ten to twenty feet *average* depth would have prohibited plant or animal evolvement.

Furthermore, Mars is now devoid of water and it is obvious that it should have some water cover, since the planets Jupiter, Saturn, Neptune and Uranus have progressively deeper water covers. The mode of birth of the planets governed the increasing volume of water on each planet from Mercury to Uranus. Therefore, something drastic happened that reversed the order of the original water cover of at least four of the planets. Earth lost its water to Saturn (rings). Mars lost its water to Earth and Pluto lost its ocean to Jupiter and Saturn – possibly some of it traveled further into deep space.

And *strangely* just enough water arrived on Earth from Mars to flow slightly above the edge of all the continental shelves.

The table on page 60, Fig. 51, gives the approximate original water volume that was extracted from the Sun and accompanied each planet into space when they were created. Water, when extracted from the Sun, was in the form of hydrogen and oxygen. The atmosphere of Earth has a much higher oxygen content than formerly, when nitrogen was high and oxygen was lower. The present atmosphere is new to Earth and came from a planet that was created prior to the Earth. The ratios of oxygen, nitrogen and other gases were of different consistency (lighter elemental types) than that which accompanied the Earth when it was created, since the Earth was extracted from deeper within the Sun's cindershroud. The ratios of gases on the planets followed the same rules of extraction from the Sun as water and the other elements.

The planet Mars was converted from a lush, rich body teaming with life into a total wasteland. Only faint traces of some forms of plants will be found to still exist on Mars. The atmosphere is extremely thin and very small amounts of water exist in its crust.

Nearly all of the original marine life on Earth was destroyed during the catastrophe. A small amount remained. Plants in the tropic zones, and on each side for hundreds of miles, were totally destroyed, broken up, ripped out and buried. Thus, the great ocean transfer was accomplished in a very short period of time – a matter of a few days – replacing the losses on one planet while totally robbing another.

Special note: Had the Viking 1 lander been placed on the continental crust of Mars instead of in the dry ocean floor, forms of life would have been discovered.

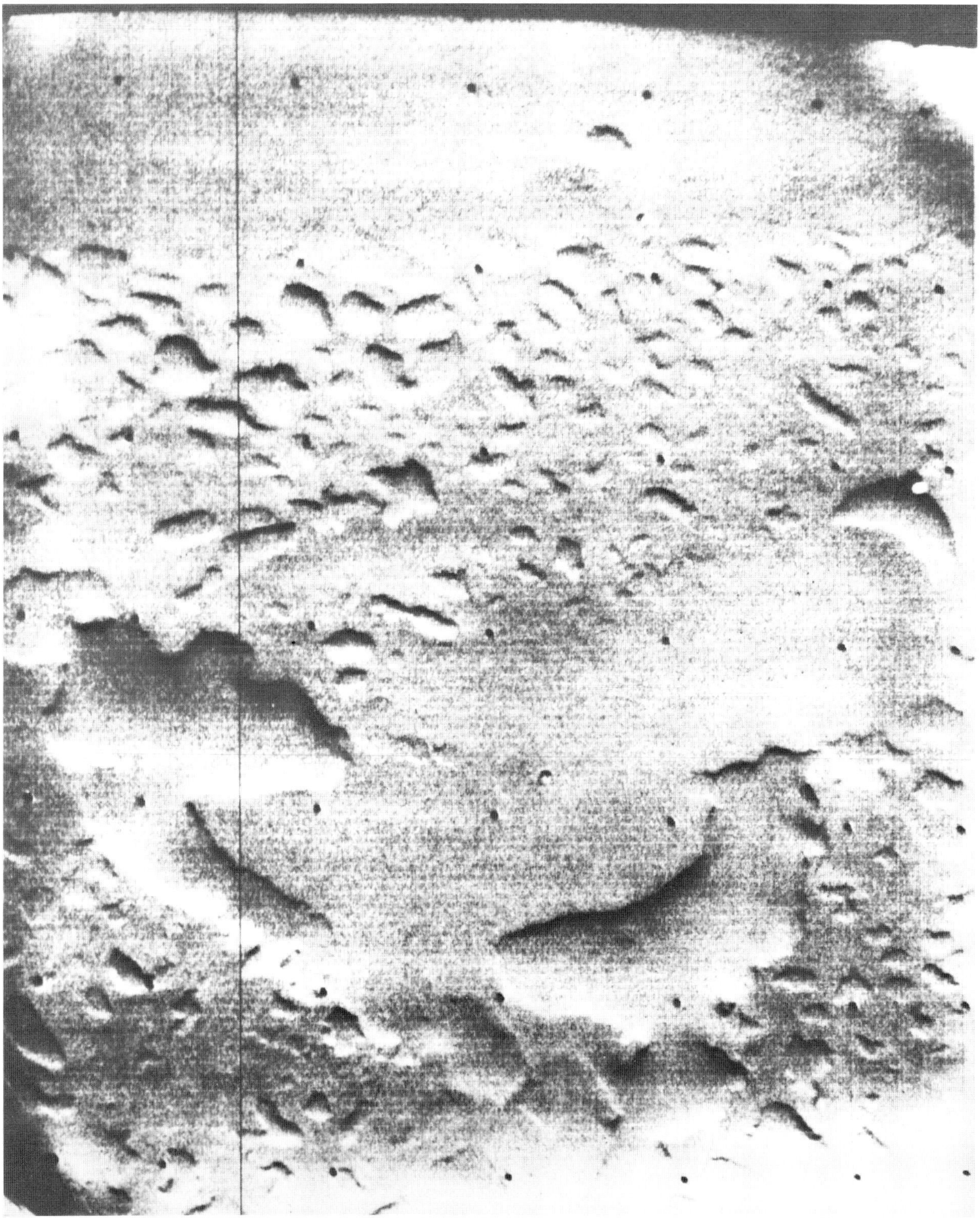

Fig. 182. Some "unusual" Martian features. NASA

(NOTE: Compare Earth's Great Lakes area without water.)

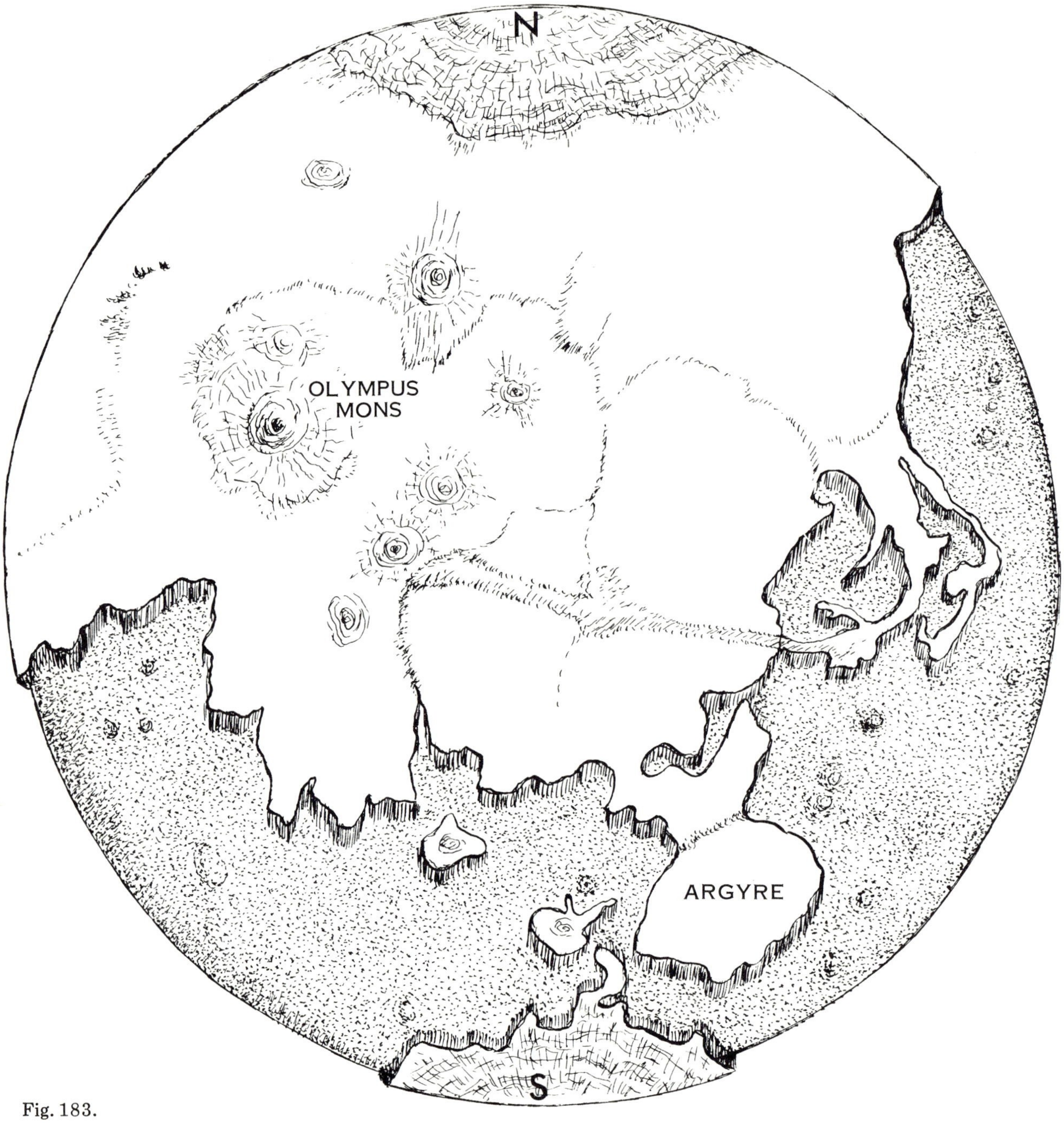

Fig. 183.

The above drawing of Mars shows a striking resemblance to the Earth with its water removed. On Mars the "continents" stand above the great empty "ocean basins" now dry and dotted by craters. Many islands from the continents to the South Pole are scattered on the lower part of the hemisphere. Argyre is an isolated continental segment which drifted south over fluid rock before it hardened. The islands and Argyre were attempting to follow the greater mass of crust that departed to become the moons of Mars and portions of the asteroid belt.

In the right hemisphere a great segment of the continental crust known as Hellas broke away

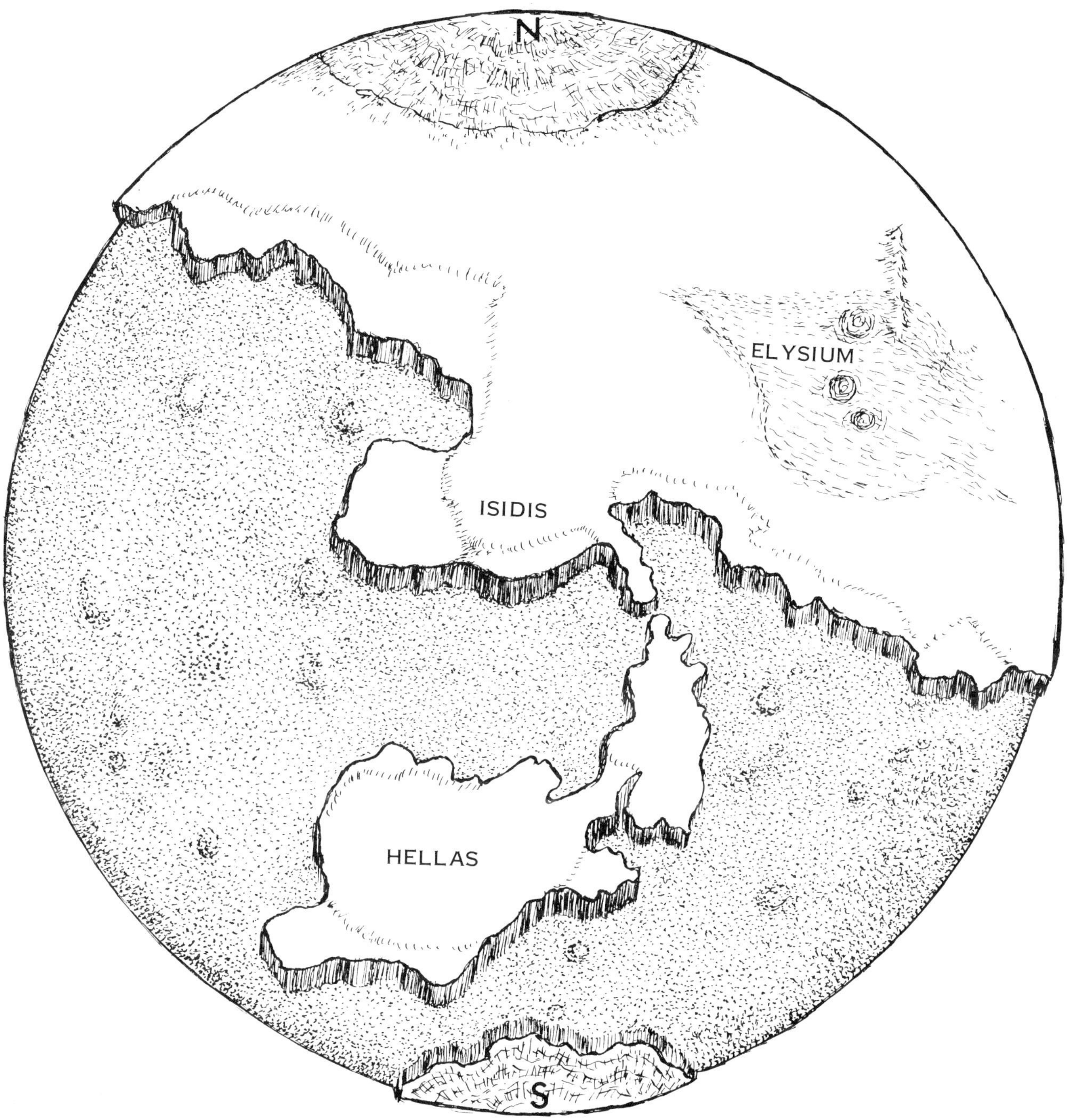

from the main part of the remaining crust along the "coastline" of what is known as Isidis, as did Australia when it broke away from the African and Persian coast on Earth.

The continental blocks and islands on Mars stand above the empty ocean floors up to three miles in stark relief. All over Mars the flow lines of water that departed at high speed are evident everywhere. Great, extremely deep canyons exist where crustal distention took place. They remain deep because erosion by water has not existed on Mars since the catastrophe, which would have carried soil and other debris to fill them as water has filled every deep fracture on Earth with mud, sand and gravel.

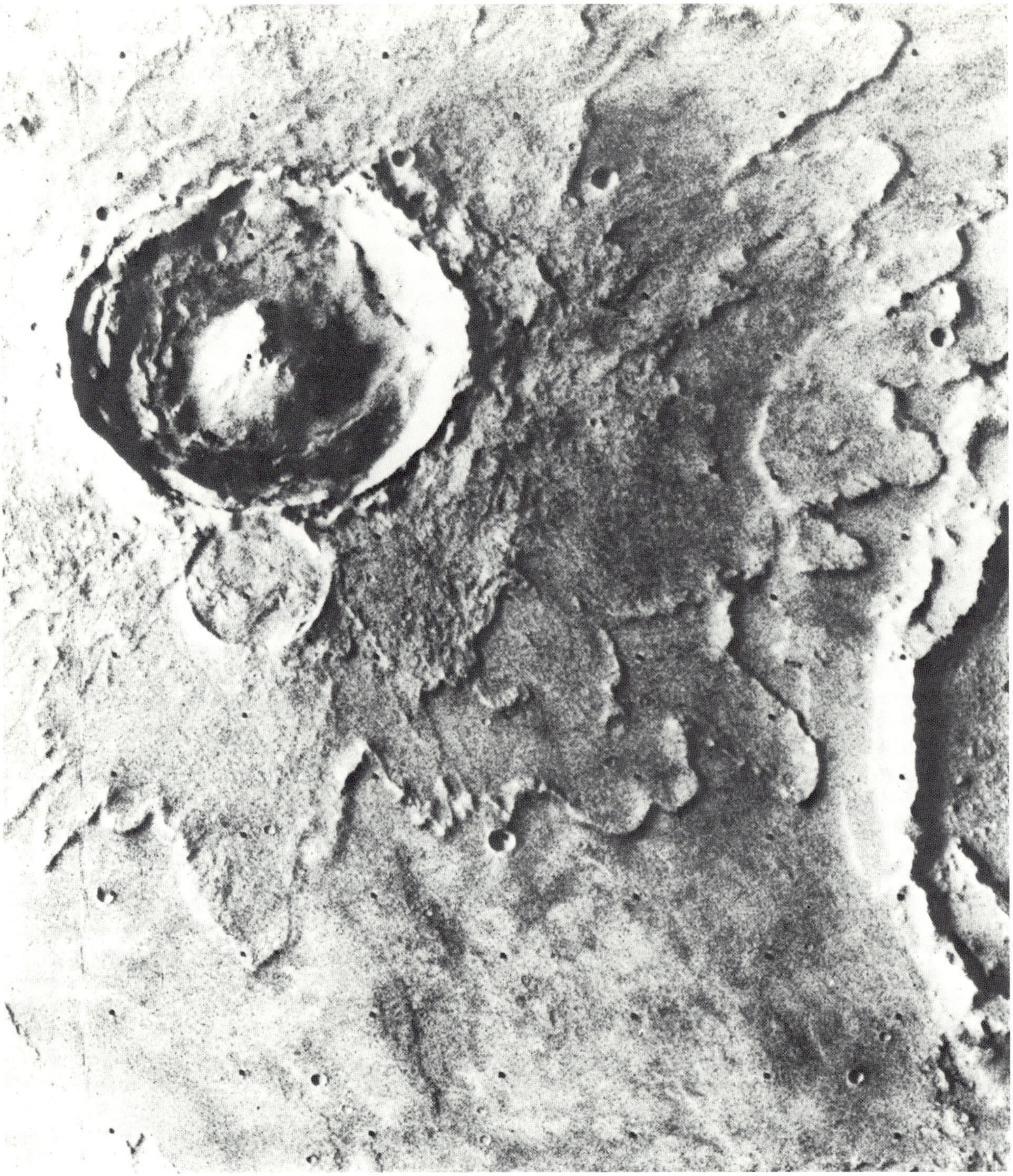

Fig. 184. Impact of muds consisting of high water contents. Notice the numerous flow lines of water as it rained away.

NASA

Fig. 185. Impact consisting totally of soft muds. NASA

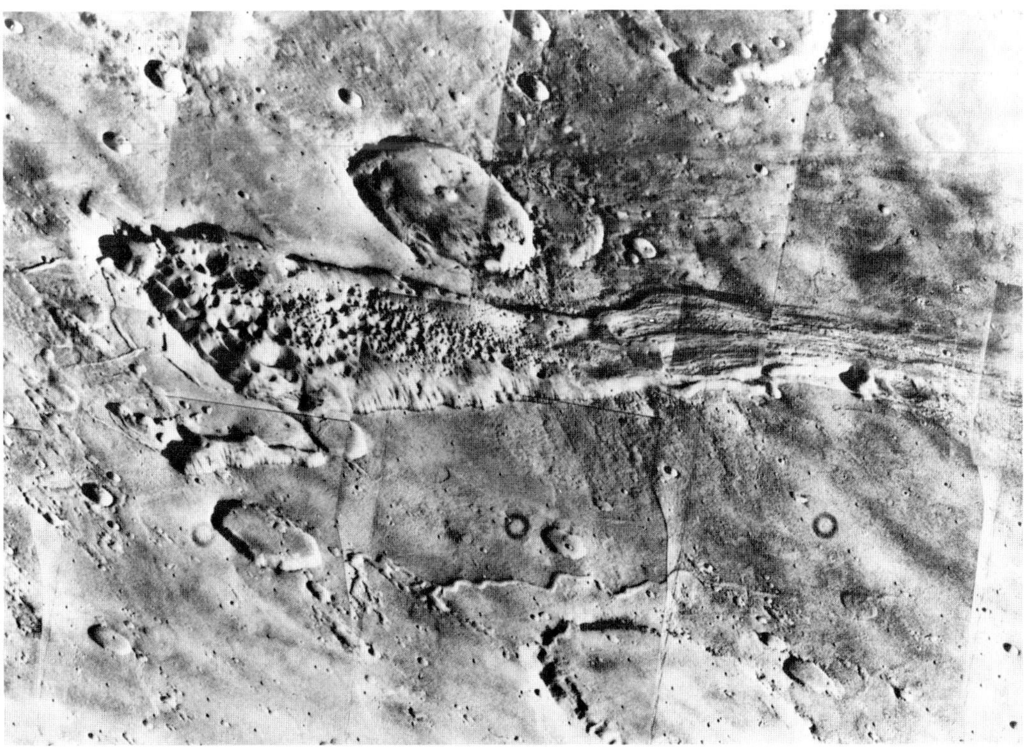

NASA

Fig. 186. A canyon fracture which partically filled with surrounding surface muds during catastrophe.

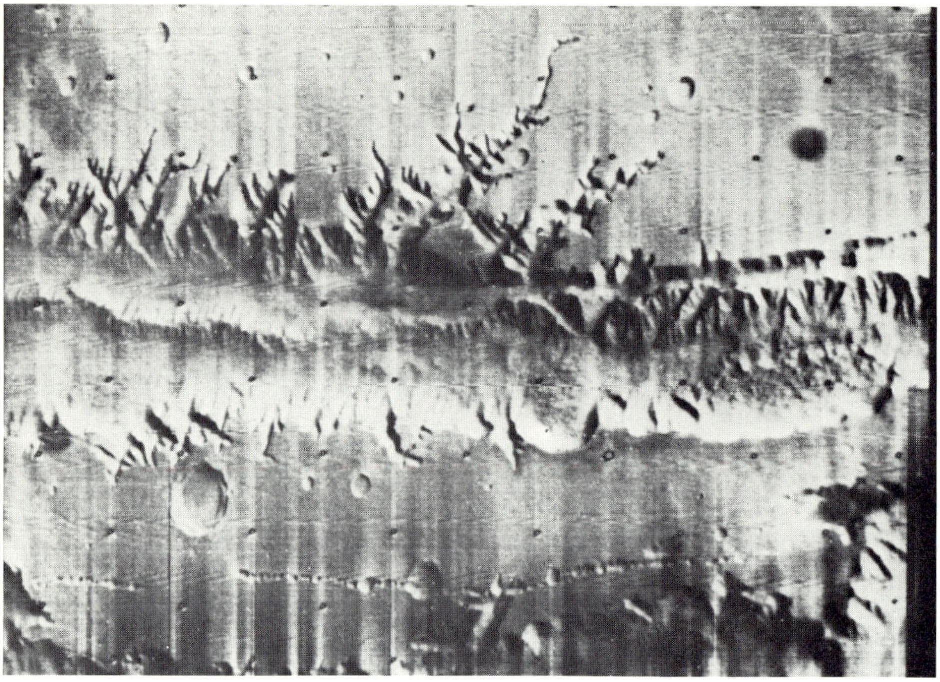

NOTE: These photographs show massive movements of water over the surface of Mars.

Fig. 187. NASA

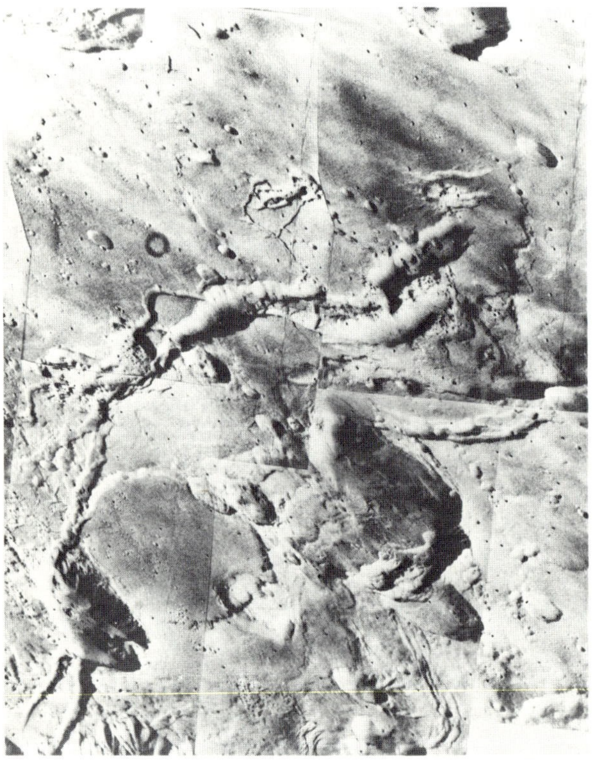

Fig. 188. NASA

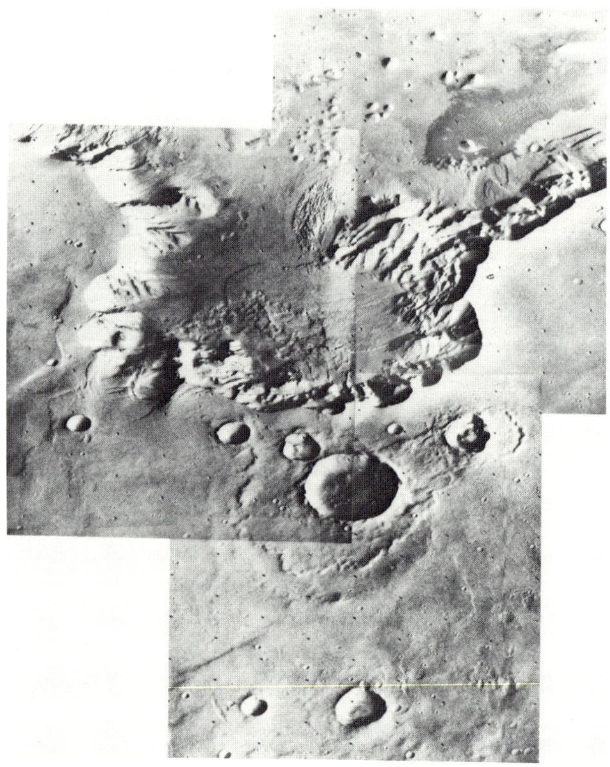

Fig. 189. NASA

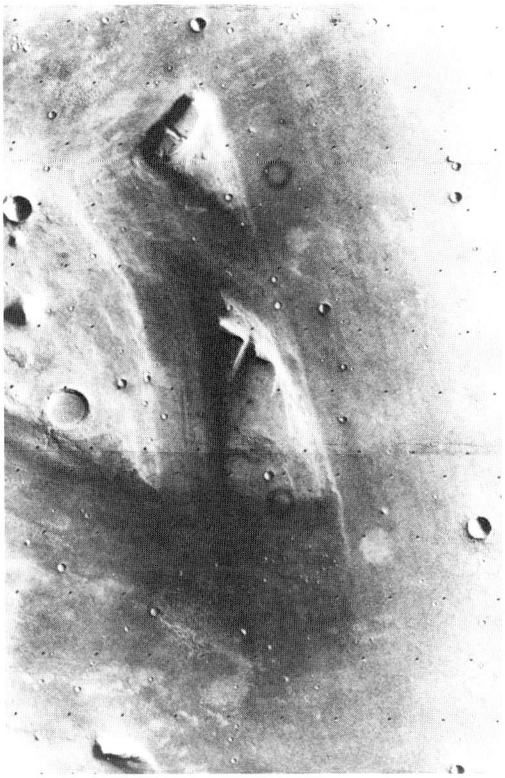

Fig. 190. NASA

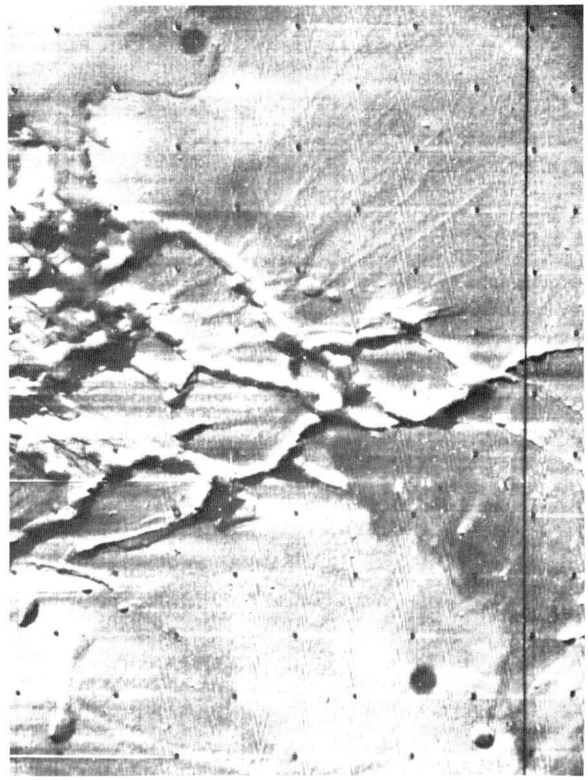

Fig. 191. NASA

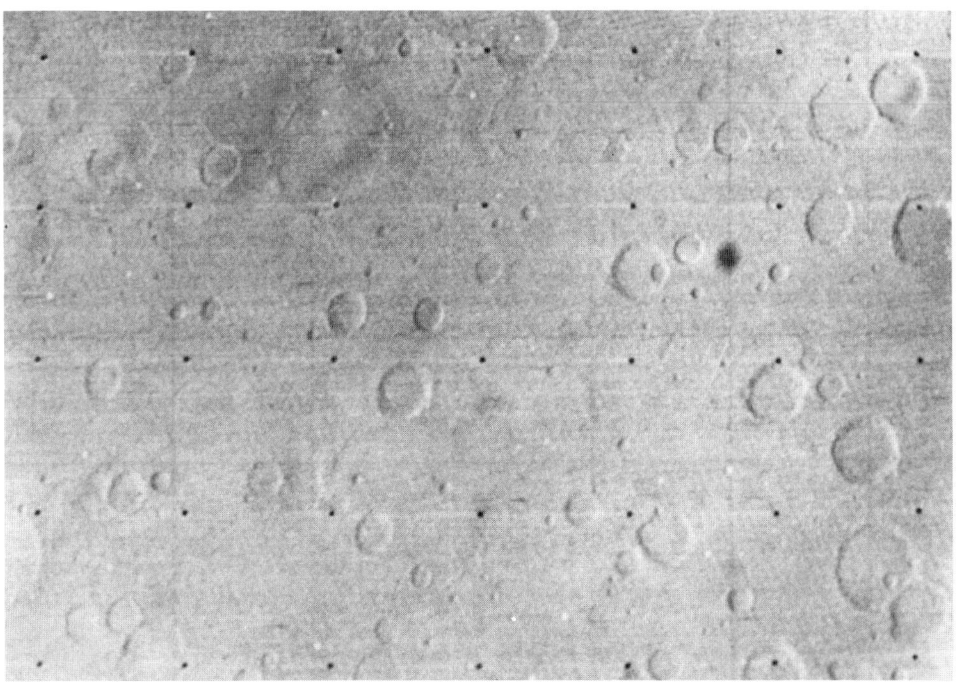

NASA

Fig. 192. All craters in the ocean basins are very shallow and sharply defined since the impacting material landed upon the hot fluid rock.

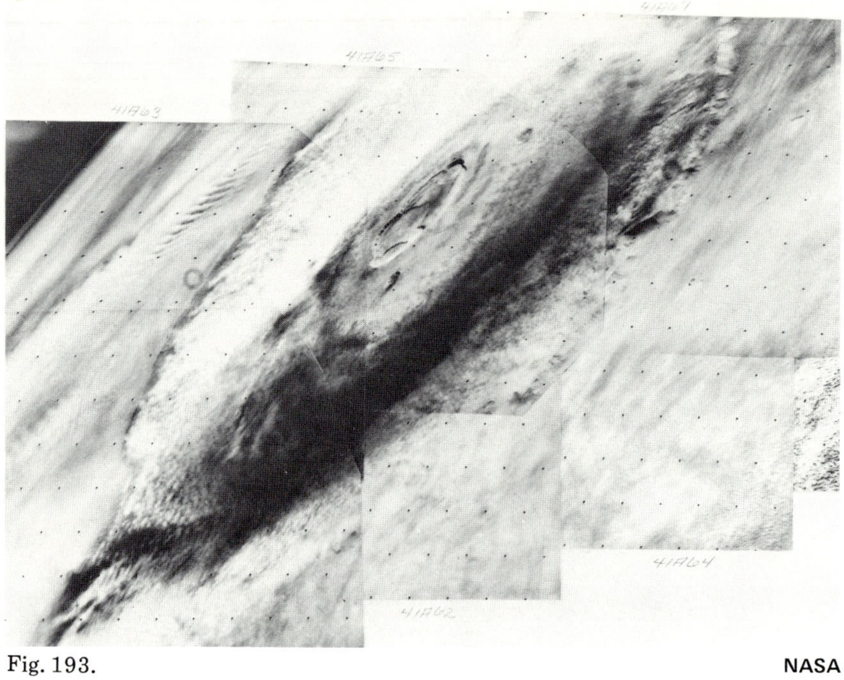

Fig. 193. NASA

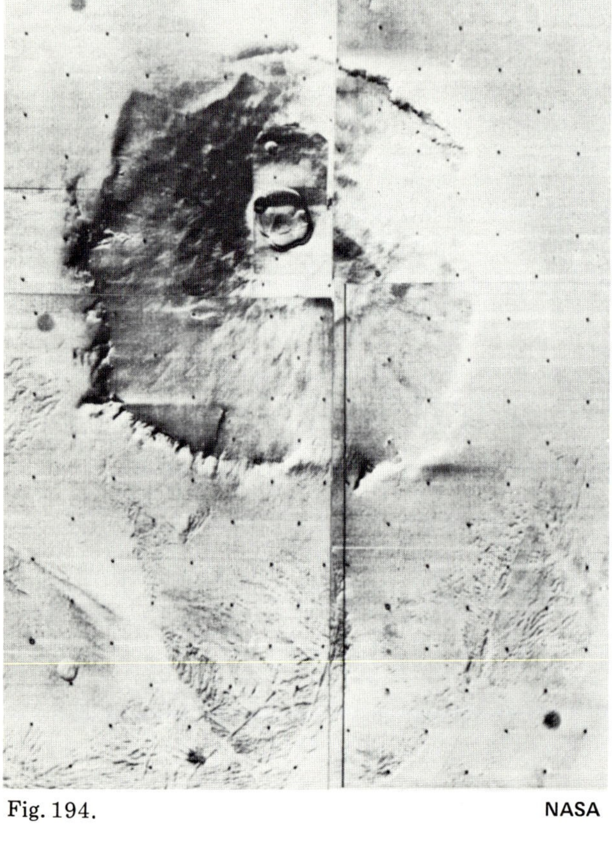

Fig. 194. NASA

OLYMPUS MONS

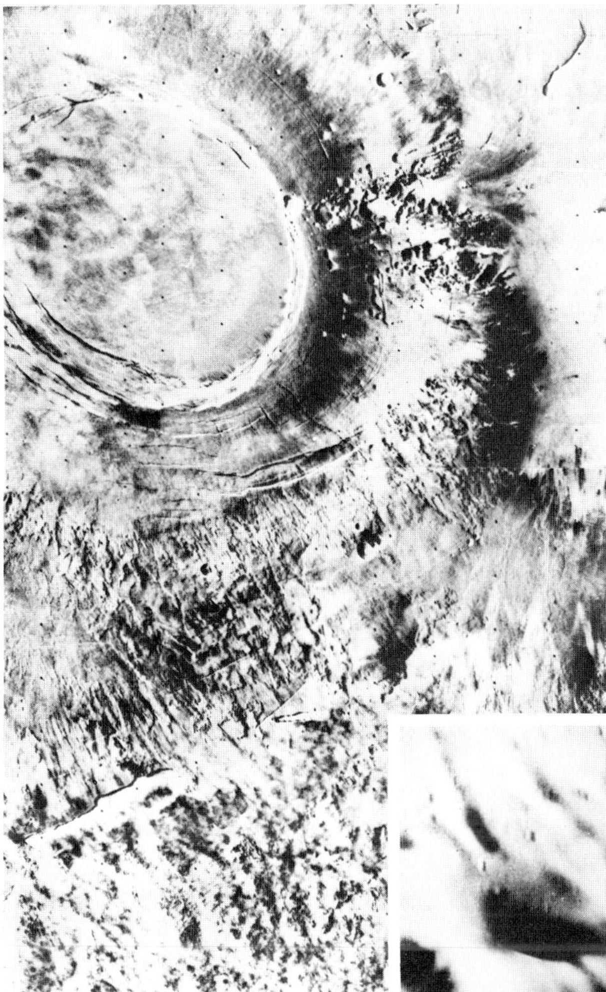

Fig. 195. Arsia Mons NASA

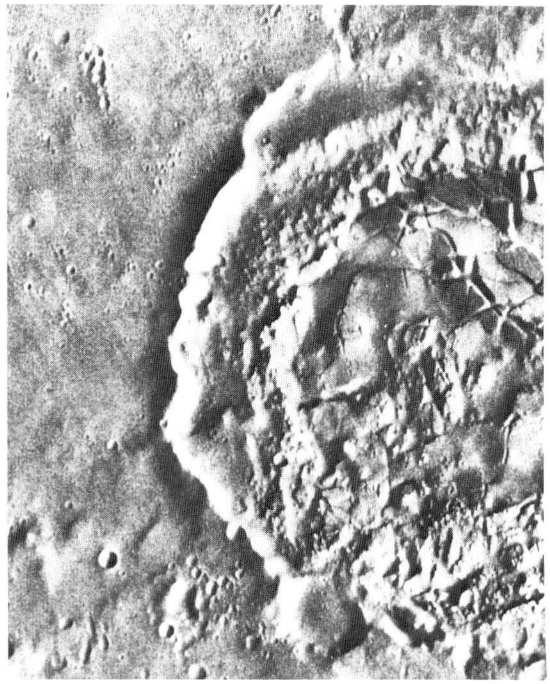

Fig. 196. NASA

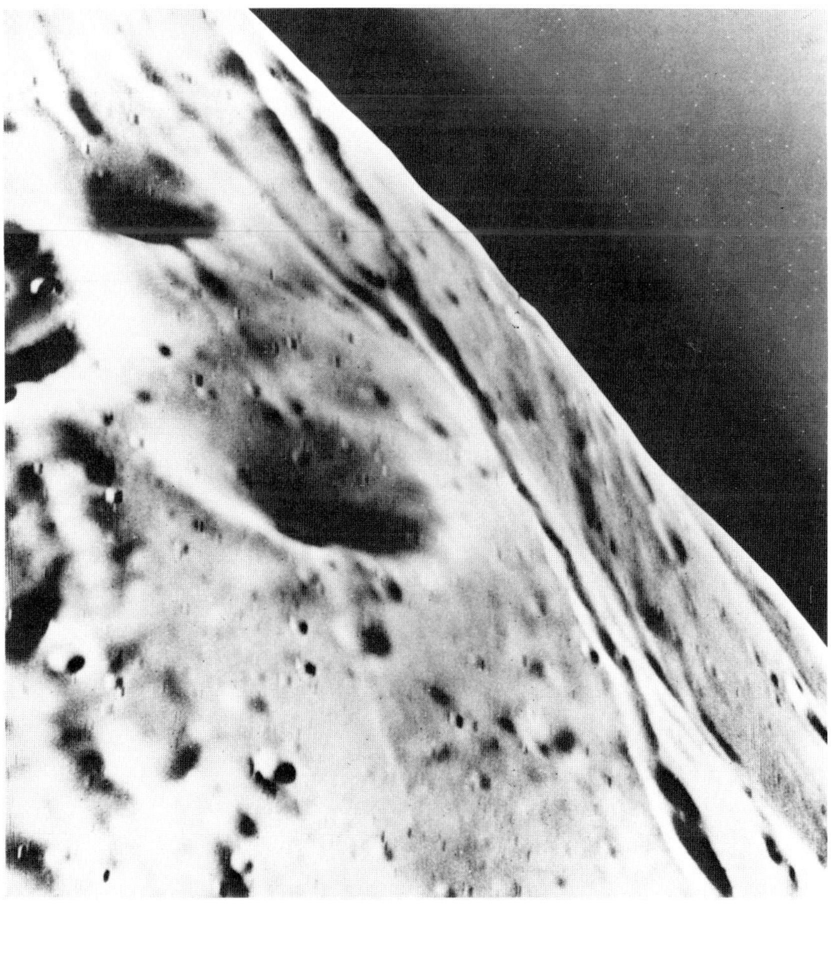

Fig. 197. Phobos

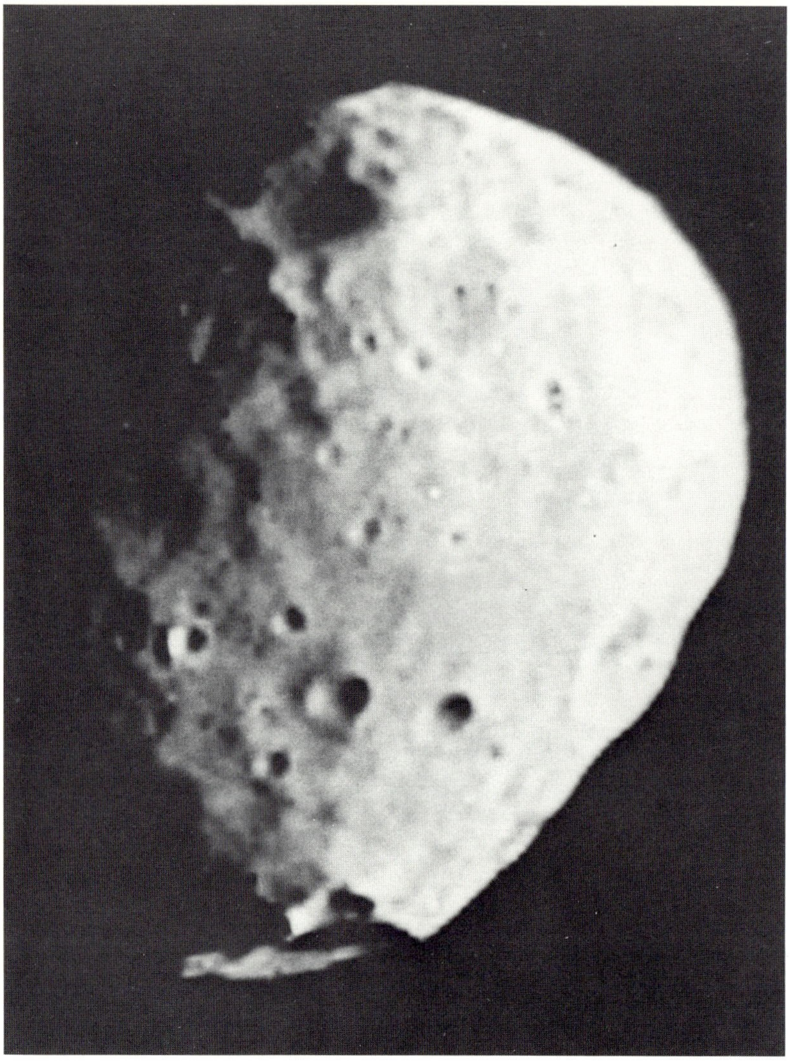

Fig. 198. Phobos was too small to form a sphere. NASA

Fig. 199. Surface of Mars from Viking 2. NASA

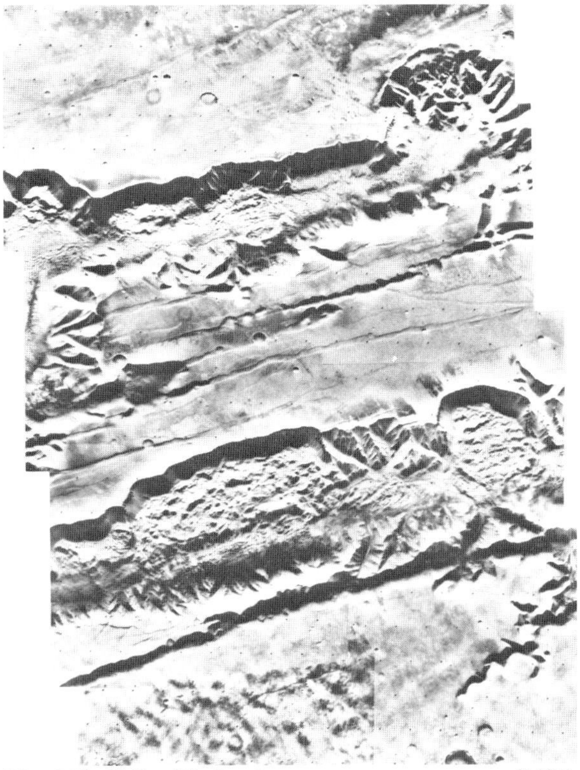

Fig. 200. Valles Marineris NASA

Fig. 201. Valles Marineris NASA

Fig. 202. Mars' Nilosyrtis region. NASA

THE UNITED STATES SPACE PROGRAM

This final chapter concerns the great attempts by man in a gigantic organized effort to learn about deep space and neighbors, the Moon, the planets and the Sun. I will be brief since the public has been made aware of much of these efforts and findings.

The Moon Probe told us much about the lunar soils, gravity, presence of water and atmosphere – and existence of former life. The manned landings of the Apollo series allowed the return of soil and mineral specimens, as well as the placement of instruments to measure various events and conditions. There were also numerous on-site studies of the Moon's physical makeup.

What was learned by this effort?

That volcanoes do not exist on the Moon.
That something had impacted the Moon in great numbers.
That there are massive concentrations of something on the Earth side that have unusual gravity anomalies.
That the maria are great basins of igneous material.
That the Sun may not be structured as presently believed.

Instead of volcanic craters, *impact* craters were found.
Metal deposits of great size.
Igneous rock of very deep origin.
Unusual forms of fused glass spheres of great beauty scattered over the Moon's surface.
Great catastrophic evidence.
Moisture on the Moon.
A very strong, steady solar wind composed of dust particles driven by the Sun.
Extreme radiation composed of cosmic rays and X rays.
Magnetic fields exist in deep space.
Temperature of deep space not as "cold" as formerly thought.

CONCLUSION

Based upon the information presented in this book, we are faced with many new concepts of the Earth, the Moon, the planets and the Sun. Yet, using such information, we are now in a position to examine more closely the true function, not only of the Sun and the planets, but the universe as well. For without the structures of the Sun and the Earth as presented here as a basis, the activities described in this book could not have happened. Current theories as proposed by most authorities fail to answer many of the most important questions, they only result in making the universe more complex and mystifying – simply because those proposing such theories do not have the "key."

In this book I have presented what I believe to be the key, the key that unlocks the many mysteries of our Earth, Moon, Sun and the universe. This key explains the real functions of the Sun, its true structure, and its place as parent to the planets. It explains why the Earth and the other planets are able to survive a catastrophe and readjust from the after-effects. It explains their innate stability and resiliency.

The key explains why the Pacific Ocean is different from the other oceans and why it was affected in such a strange way. Why it contains the many great fractures lacking in the other oceans, and why its floor is covered with metal.

We can now understand why the former crust of the Earth was able to drift – swiftly – into its many present positions and that it did so as a natural result of the the rebalancing forces present within the Earth. This adjustment of the crust created the continents, the islands, and the oceans, mountains, valleys, lakes and rivers.

We can now see how the Moon was born in an almost simple manner and why its surface appears as it does. We can now better understand and more accurately interpret the results of the different space and lunar probes. The surface features and objects photographed are easily identified and understood. Confirmation of many statements in this book are arriving from laboratories, observatories and scientific centers almost daily. The last Russian landing, as well as our own Surveyor III and recent Apollo landings, indicated that the soil on the Moon was similar to that of Earth.

We can now understand why we have seasons, weather, thunder, lightning, wind, rain and snow. Why great continental ice sheets could never have existed. We now know what actually caused the evidence attributed to them. Remember, the "Ice Ages" were created to answer certain puzzling questions, questions of deposits of till, loess and clay, moraines and mountain lakes, erratic and striated boulders found hundreds of miles from their sources, sharp mountain peaks and deep valleys. Then there are those acres of granite, polished smooth and gleaming – and the Canadian Shield. These many geological features are simply the outcome of the violent global and surface adjustment that followed the birth of the Moon. We can now understand why there was a sudden and massive faulting, great sections of limp crust, folding, twisting, sliding over other sections and coming to rest in jumbled heaps. We can see mountains formed in moments when great gushes of liquid rock shot into the outer crust and the air. We can now know why the Earth is covered with great beds of some of the most beautiful objects on Earth – gravel. We know where sand really came from and why the crust of the Earth is filled with billions of tons of metal of all kinds, and where this metal came from.

In all – the universe is a beautiful and complex system, yet all parts function in concert with all other parts. A theory must fit all observed phenomena, it must answer questions not pose new ones.

Fig. 203. Horsehead Nebula in Orion. **Mount Wilson and Palomar Observatories**

Fig. 204. Large crater in background is Copernicus. NASA

Fig. 205. NASA

Fig. 206. NASA

Fig. 207. NASA

Fig. 208. Crater Tsiolkovsky (a large sink) on the far side of the Moon. NASA

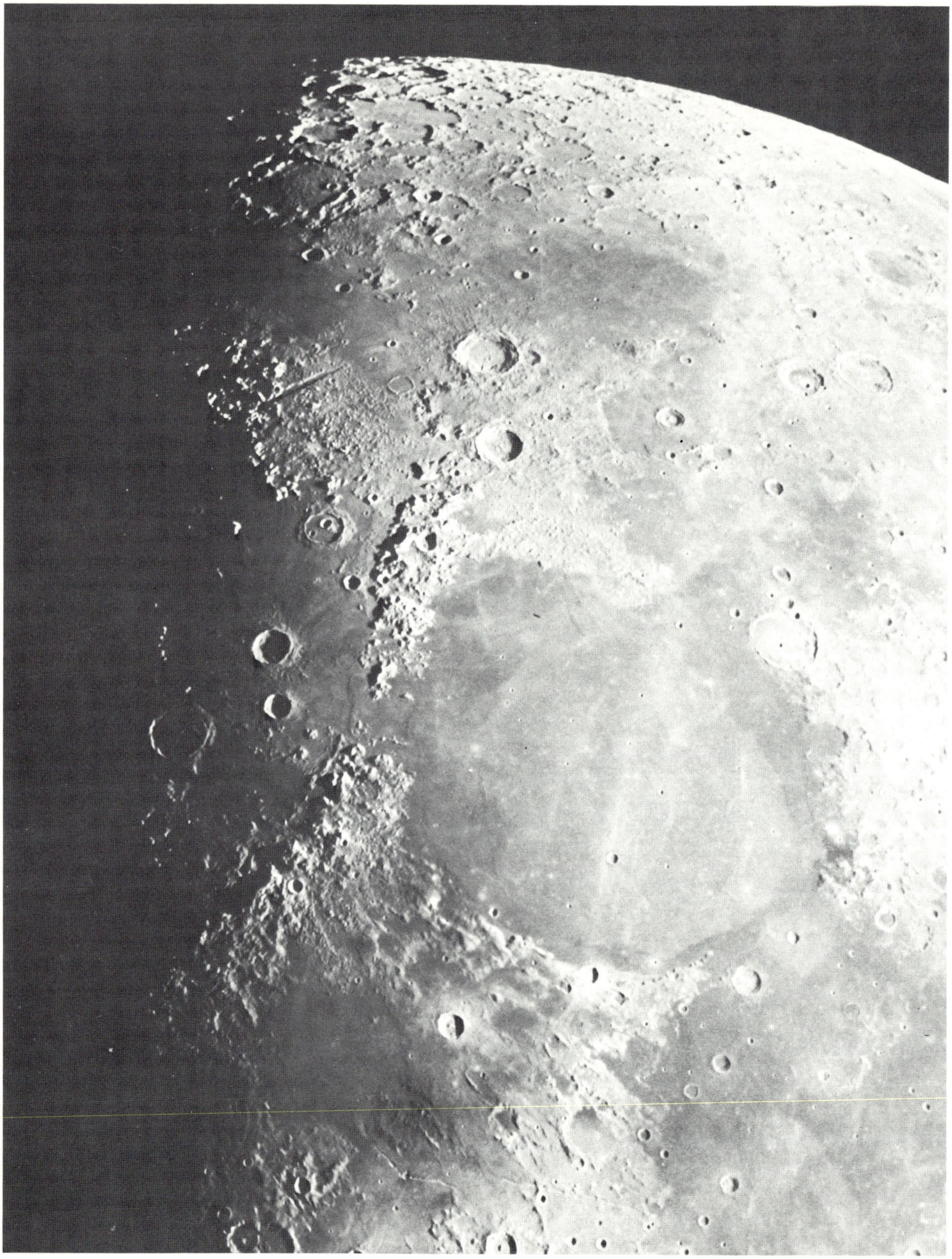

Fig. 209. Mount Wilson and Palomar Observatories

How Fossils And Gravel Were Formed

SPECIAL NOTE

This brief section on gravel is included in this book on the Moon because it is a link in the chain of events that brought new kinds of materials to the surface of the Earth. These new materials are clues to activities – sudden and universal in nature – that reveal extensive details of the Earth's interior structure that would not otherwise be available to us for investigation and study.

The author has found thousands of specimens of gravel (nuggets of the Earth), many of which are solid metal, composed of copper, gold, manganese, iron, tin. Many of the metal nuggets contain fossils of seashells, ferns, nuts and coral!

All gravel (except in stream beds) have smooth unscratched surfaces, even under a microscope no scratches are evident. If gravel had been formed by the action of glacier movement, the grinding and polishing involved would have left a completely abraded surface. On the other hand, creation of gravel from fluid materials that were shot to the surface of the Earth from deep within the Earth and cooling in transit left the gravel with naturally smooth, unabraded surfaces.

By examining under a microscope any of the specimens shown in the following pages, all cells, surface and interior structure will be seen to be present and faithfully preserved.

The manner in which the fossils and gravel shown here were created was relatively simple. As the great tidal flood was rushing over the Earth just a few days prior the Moon's extraction, millions of deep fractures began opening everywhere in the Earth's surface.

Enormous volumes of debris-laden water containing all forms of life poured down and down into the fractures to depths as much as 60 miles. The great heat within the openings volatilized the waters, leaving the debris at the bottom of the chasm, below the oxygen level. As the fault closed, liquid rock from beneath rose and engulfed the life-rich debris. As the fracture slammed shut, some of the fluid rock shot out of the ground, spurted high into the sky, then peppered down on the surface of the Earth. Most of the newly formed gravel cooled in the air before landing. Much of it was dented or deformed into different shapes on impact.

Most of this gravel is found on the surface of the Earth where it originally landed. Many contain fossils trapped within.

This stone was shot out of what is now the North Platte River while it was momentarily a deep, open chasm. The stone came out of the open fault in a fluid (plastic) state. The outer surface of the material had cooled by the time the stone landed on the ground and on other stones. The impact fractured the cooled outer crust like a smashed chocolate creme, but the still plastic interior kept the stone intact. The point of impact is easily seen. Some of the stones were still so hot that the points of impact fused with other materials they touched. The source of the above stone is alongside the North Platte River in Carbon County, Wyoming. It lay on the surface of the ground with numerous others in similar conditions. The stones range from 1 inch to 2 feet in diameter.

Same type of stone as above. It was fractured in the same manner. Several impact points indicate that it rolled and bounced while still very hot and soft inside. Note the offset where the crust tended to slip over the still plastic center.

Same type of stones as those on previous page. These stones came from the Little Colorado River near St. Johns, Arizona. They were ejected from the fault that formed the river bed.

These stones could not have been left in these conditions without still plastic centers inside a hardened crust.

This stone was produced by the same activity, but it came from central Japan. The kind of material composing all these rocks — whether found in Japan or Arizona — is different, since they arrive from different depths in the Earth. The specific gravity is also different with mineral content ranging from calcium, silicas and others to many kinds of metals.

Petrified fleshy parts of mollusk and shell in flint.

On the left is a shell imprint in sedimentary rock of the Late Cretaceous. On the right is a shell of the same kind encased in flint. The one on the left is from northern Chihuahua, Mexico and the one on the right is from the eastern edge of the Grand Canyon, Arizona.

Bryozoans trapped in flint in the same manner as previously described. Note the faithful and distinct petrification.

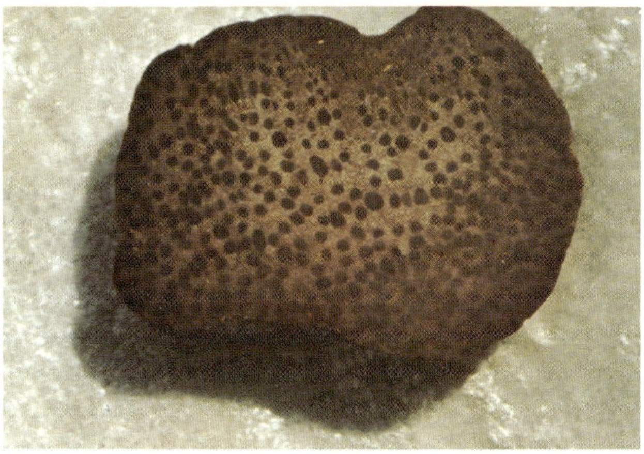

Section of wood from a palm petrified in flint. It came from the Arizona desert near the Colorado River.

Bryozoa encased within flint.

End imprint of crinoid in flint. The crinoid left a casting of its shape in the larger stone before remaining gases ejected it during petrification as the material rose out of a nearby fault.

Notice how the surface cooled, it appears glazed. In the center near the bottom is the petrified fleshy part of a mollusk.

Fleshy parts (mollusks) which were petrified almost instantly at very high temperatures.

Chert (flint) gravel containing petrified sea shells with the fleshy parts petrified within — all encased within a rock that has a melting point of over 2000 degrees F. Petrification was done in the absence of oxygen, therefore the fleshy parts and shells did not char in the heat. Only live things were petrified.

Remains of a sea shell. Pressure of residual steam blew off the fleshy parts leaving the cavity and shell.

Seashells in flint. Process of formation was same as described above.

A perfectly petrified prune-like fruit with seed inside. Now solid flint.

On the left is a Brazil nut as grown today, and on the right is a petrified Brazil nut from the Arizona Desert. Many kinds of nuts, seeds, beans are found petrified in this way around the world.

A fragment of a coconut, now petrified. One of the most important finds made since it confirms the mode of petrification. No nut in the whole world grows in the same manner as the coconut. Note the thin shell and the coconut "meat." Under a microscope the fibres are perfectely preserved.

A frog petrified in iron and rare earths. Specimen from central Texas.

Bryozoan imprints in flint. Please keep in mind that flint is not water soluble!

Pyrite encased shellfish. Specimen came from western Nebraska.

On the left is an oyster petrified by manganese. The specimen was found in Wyoming. On the right is a very large leaf petrified by hot silica gases.

An ancient fruit.

Nuts encased in iron ore. These nuts reached the Second Intermetallic Layer (about 110 miles down).

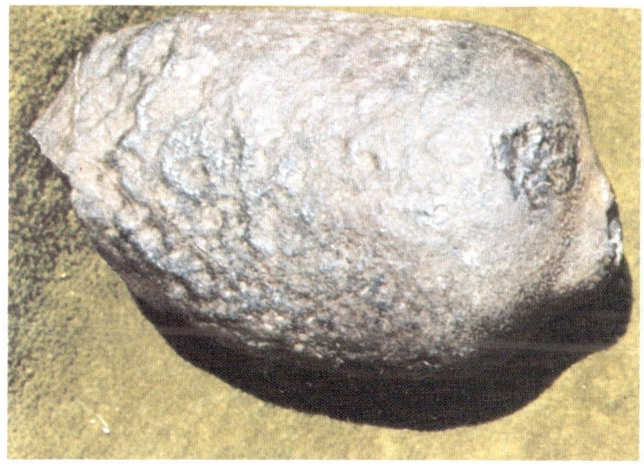

Petrified mango-like fruit. This specimen is composed predominately of iron from western Chihuahua, Mexico. It was petrified in the same manner as all others shown in this gravel section.

A shellfish completely encased within flint.

A piece of charred wood encased in manganese which came out of a nearby fracture during the catastrophe. Source of specimen: 45 miles south of Presido, Texas.

Iron ore from a mine in eastern Chihuahua, Mexico in which asphalt is enclosed.

Geode containing crude oil! Specimen from central Illinois.

Geode from western Chihuahua, Mexico in which pure gold as a gas condensed on the quartz crystals. Depth of origin: top of Second Intermetallic Layer, about 90 miles down;

Cavity in geode caused by hot expanding gas. Found in Montana.

Geode with pure quartz lining (Mexico).

Slice of quartz containing pure copper. The copper originated from the First Intermetallic Layer, a depth of 490 miles. Specimen came from eastern Pennslyvania.

Geode showing interior crystals.

Nuggets of the Earth. Bottom left: pure gold in chert gravel from Montana. Top left: pure manganese and silver from Sweetwater County, Wyoming. Right: tin nugget from area north of Aquas Caliente, Mexico. Center: pure gold nugget from Alisios, Chihuahua, Mexico.

Mammoth bones petrified in flint. (Western Wyoming)

Concretion

Sea life fossilized by fluorspar gasses.

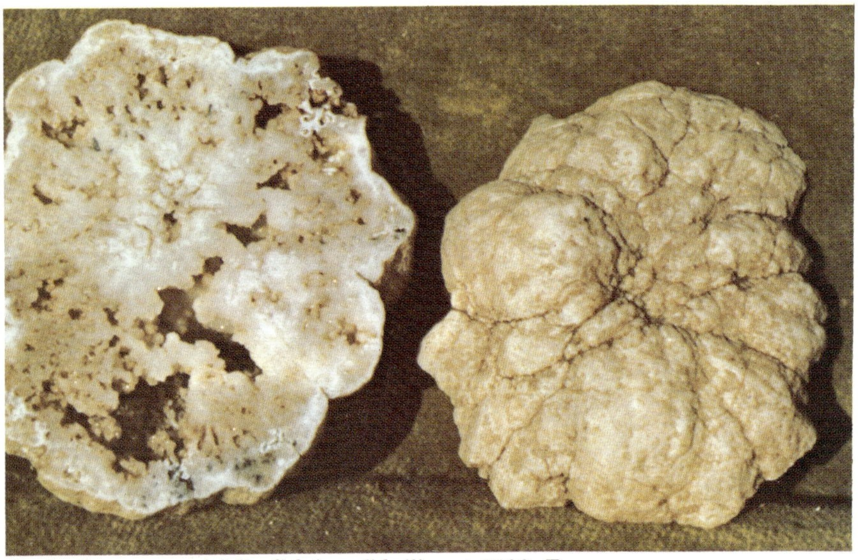

Perfectly preserved jelly fish in solid silica at 2500 F. Specimen from a gravel pit in west central Indiana. Under a microscope all details are evident.

Petrified sponge.

Wood impregnated by copper on left and by uranium on right. Both elements enveloped the wood as hot dense gases.

APPENDIX

REASONS WHY THE DATE OF THE CATASTROPHE AND BIRTH OF THE MOON WAS VERY RECENT

1. Carbon dating of the age of petroleum does not average much more than 8,000 years.
2. The ages of volcanoes are not much over 6,000 years and they are diminishing in activity worldwide because the very deep petroleum reserves that feed them are nearly depleted.
3. Deltas of the major rivers, worldwide, are not much over 10,000 years old.
4. Carbon dating of the morass of death around the arctic circle shows a date less than 8,000 years..
5. New sediments on the ocean floors are but a few inches deep.
6. The *new* histories of man (all ancient materials uncovered in the past few hundred years) do not predate 7,000 years.
7. Depth of dust on the various objects on the Moon atop debris scattered over the surface is less than one and a half inches deep. (Dust from the Sun builds one inch on the Earth every 10,000 years – at present.)
8. Beach debris on the shores of all the oceans is very shallow in depth – sea shells, sand, old sea weed, coral, gravel. This indicates a very recent (new) formation.
9. Most important of all is the new forms of life on Earth – recent arrivals which have no prior evolvement on Earth – flora or fauna. What little "fossil" evidence remains, lies within two to six inches of the present surface. Most new life arrived from Mars, with some from Venus.
10. The new forms of life on Earth which have no prior evidence of evolvement on Earth include photo-synthesizing algae, calcareous foraminifera.
11. Gravels lying all over the top of the ground in the painted desert of Arizona – all containing fossils of Late Cretaceous life – *these are recent;* otherwise, why hasn't eroding forces or "glaciation" removed this evidence millions of years ago? All other gravel found atop mesas, plateaus, and other exposed areas worldwide whose appearance and origin is unexplained. (See Appendix: "How Fossils and Gravel Were Formed."
12. The petrified trees of the Cretaceous are also lying on the surface of the Petrified Forest, as are all others around the world. And why are petrified roots of trees never found?
13. The gold nuggets found lying on and in the surface soils of many parts of the world. I have found copper, manganese, lead, tin, platinum, tungsten, gold and silver nuggets lying *on top* of the ground in many parts of the world. *These are of recent origin;* otherwise, they would have been flushed off the surface of the ground long ago or been buried.
14. The oceans floors are *new.*
15. When the coral atolls in the Pacific Ocean are measured – how fast do they grow? How long would it take to reach their present heights? The answer is: 6,000 to 10,000 years.
16. Salt domes – massive things – found *only* where oil is found and *only* in certain areas of the world.

17. Fossils in perfect condition can be found lying in surface gravel deposits near faults in Illinois, Indiana, Ohio and Canada. These are petrified jelly fish, sponges, plants, oysters. I have found these and many more. Enormous volumes of these are in the areas mentioned. The same kinds are found at the Equator. All these fossils are mixed in with gravels, materials of which came from deep in the Earth, some 300 to 400 miles down. Some of these gravels contain metals. (See "How Fossils and Gravel Were Formed."

18. All geodes are found lying on the surface of the ground. Many contain metals in crystalline form coating the interiors of the geodes. Some have pure gold inclusions in the center (Pg. 190) with hundreds of specks mixed with the rock of the exterior. In one area, carnotite crystals fill the void of the geodes. These have been found along the Rio Grande on the border of Mexico and Texas on the Texas side. All geodes are found near a major fault. This applies anywhere in the world. They,too, are of recent origin. All geodes originate from depths of 50 to 300 miles in the Earth.

19. The injections of metals in veins in the rocks worldwide is recent. The metals in the veins have not had time to be eroded or to oxidize, indicating recent arrival.

20. Geodes with fossil inclusions are numerous. I have many examples of them. Geodes are composed of chalcedony (agate), all a form of quartz which is basically silica that has been formed under very high temperature and shot up through a fault in the Earth to the surface as the fault closed. At their point of origin, each geode-to-be contained a very minute gas bubble under high pressure which expanded as it neared the surface, due to reduction of external pressure which caused the gas to expand the diameter of the fluid rock mass, thus creating the geode. The gas in the center was usually composed of some mineral in a gaseous state that condensed on the internal wall of the geode as it cooled after it landed on the surface of the ground.

    Fossils trapped within geodes were alive when they were flushed into a fault and became mixed with rising liquid rock. The moisture in the organic object created steam (under high pressure) that expanded as it was ejected, creating another class of geode.

21. The ruins of many ancient cities and villages, worldwide, buried in Cenozoic (Neozoic) debris; such as Copan, Karnak, Tikal, Teotihuacan, some ancient Grecian cities, Phoenicia, Machu Pichu for example (not to be confused with volcanic debris, such as Pompeii was buried in).

22. All caves around the world are *new.*

23. Sedimentation of the floors of the major lakes, worldwide, is new and scanty. The floors of Hudson Bay, the Great Lakes, Salt Lake, Agazziz, are *new.*

* Distance from Earth: 221,000 miles to 235,000 miles
* Diameter: 2160 miles (¼ that of Earth)
* Circumference: 6785 miles
* Density: 3.3 (water = 1) (3/5 that of Earth)
* Mass: 81 trillion million tons (1/81 that of Earth)
* Volume: 5,290,790,400 cubic miles (1/49 that of Earth)
* Surface area: 14,655,600 square miles (1/14) that of Earth
* Visible surface: 59%
* Temperature: at noon, sub-solar point (radial center) 212 degrees F
at midnight, -238 degrees F.
* Albedo: 7% (1/400,000 as bright as sunlight)
* Inclination of orbit to eliptic: 5 degrees, 08 minutes, 43 seconds
* Surface gravity: 1/6 that of Earth
* Escape velocity: 1.5 miles/sec.
* Velocity of Moon around Earth: 2,287 miles an hour (in one hour moves distance = one lunar diameter.
* Synodic month: 29 days, 12 hours, 44 minutes (relative to the Sun)
* Sidereal month: 27 days, 7 hours, 44 minutes (relative to the stars)

Index

Index

SPECIAL NOTE

This book, *The Birth of the Moon,* is but a chapter from a much larger work titled, *Out of the Grey Mist.* The first book to be published from it was *The Primal Energy Transverter.*

Other books in preparation and soon to be published from *Out of the Grey Mist* are:

* The Primal Energy Transverter – Second Printing
* The Living Sun
* Lightning-Weather - Climate
* Mineral Treasures of the Earth - How and Where to Find Them
* Why the Earth Quakes
* The Structure and Physiology of the Earth
* The Creation of Petroleum
* Volcanoes
* Formation of the Continents
* The Tectonic Moment - The Creation of Mountains
* Gravel - The Nuggets of the Earth
* The Great Ocean Transfer
* Gravity
* The Structure and Cycle Mechanics of Stars

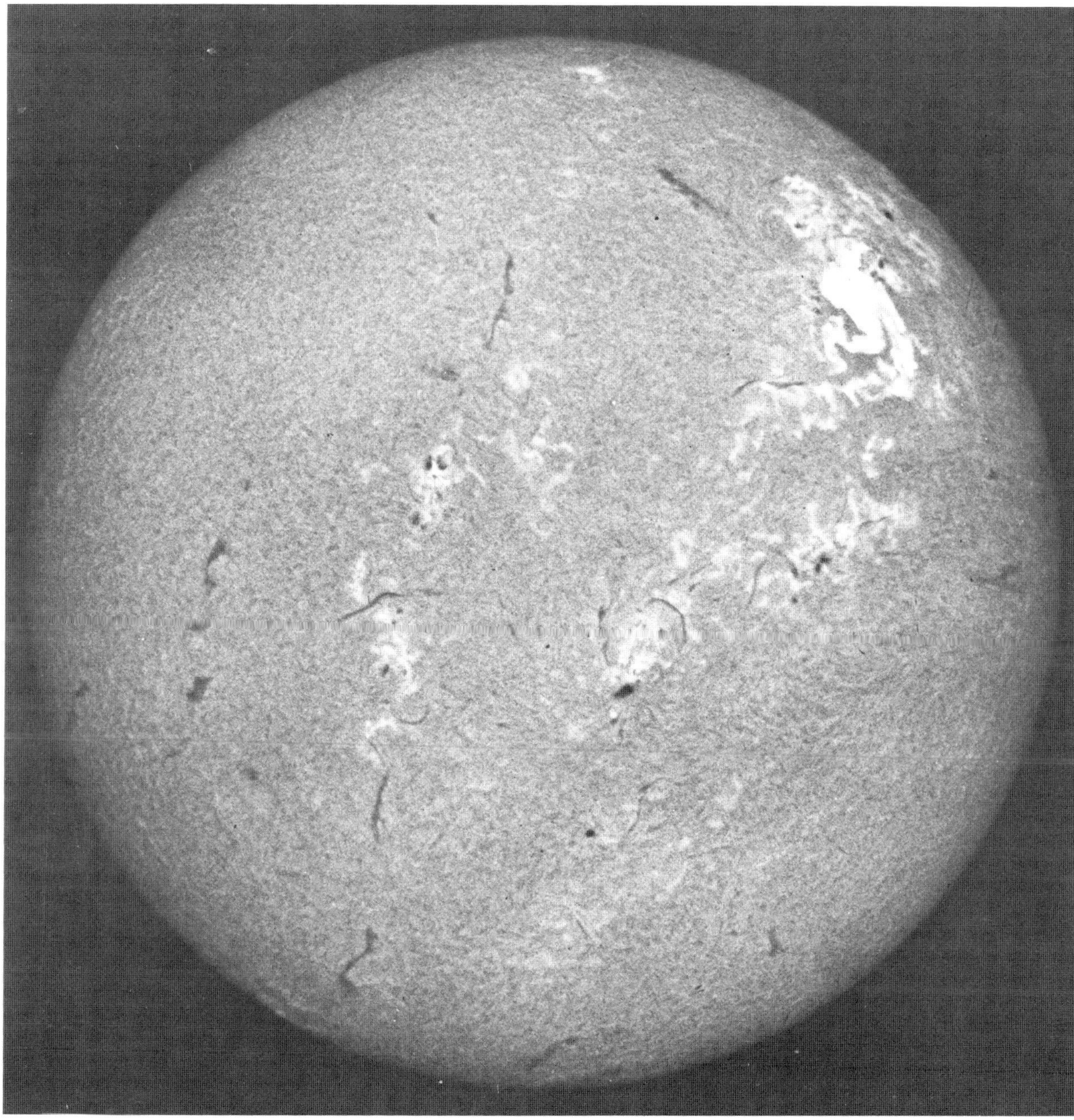

Fig. 210. Lockheed Solar Observatory

Fig. 211. NASA

Fig. 212. Lick Observatory

ABOUT NORTHROP UNIVERSITY

Northrop University, located in Inglewood/Los Angeles, California, is an accredited university through membership in the Western College Association, and is accredited as a senior institution by the Western Association of schools and colleges and is a member of the Association of Independent California Colleges and Universities. It is chartered to bestow degrees ranging from Associate to Juris Doctor degrees.

The University's history since its establishment in 1946, has placed emphasis on advanced technologies, specifically in engineering and the physical sciences. This background provides the basis for the multi-disciplined burgeoning field of energy education, analysis and applications.